THE LIBRARY
COLBY JUNIOR COLLEGE
COLBY JUNIOR COLLEGE FOR WOMEN
PARATI SERVIRE
MENS ANIMUS
CORPUS
1837

ENZYME
ANTIGEN AND VIRUS

ENZYME
ANTIGEN AND VIRUS

A Study of Macromolecular Pattern in Action

BY

F. MACFARLANE BURNET

Kt., F.R.S., F.R.C.P.

CAMBRIDGE
AT THE UNIVERSITY PRESS
1956

PUBLISHED BY
THE SYNDICS OF THE CAMBRIDGE UNIVERSITY PRESS
London Office: Bentley House, N.W. 1
American Branch: New York
Agents for Canada, India, and Pakistan: Macmillan

Printed in Great Britain by
Spottiswoode, Ballantyne & Co. Ltd.
London and Colchester

CONTENTS

Preface *page* vii

I Introduction: Enzyme Action and Protein Synthesis 1

(1) Enzyme specificity, *p.* 5. (2) Adaptive enzymes in micro-organisms, *p.* 10. (3) Chemical aspects of the biosynthesis of protein, *p.* 17. (4) The nature of adaptive enzyme synthesis, *p.* 31.

II Antibody Production 40

(1) The self-marker concept, *p.* 40. (2) Antibody production after the elimination of antigen, *p.* 44. (3) The site of antibody production, *p.* 49. (4) Theoretical approach to antibody production, *p.* 53. (5) Weaknesses of the present hypothesis, *p.* 76.

III The Self-marker Hypothesis in relation to Cellular Proliferation and Control 81

(1) Immunological aspects of tumour transplantation, *p.* 81. (2) The implications of cutaneous sensitization to simple compounds, *p.* 89. (3) Application of Weiss's concepts of cell control to the self-marker hypothesis, *p.* 94. (4) Summary, *p.* 106.

IV Virus Multiplication 109

INFLUENZA VIRUS MULTIPLICATION: 113

(1) Nucleic acid in relation to influenza virus, *p.* 116. (2) An attempted visualization of the structure of influenza virus particles, *p.* 119. (3) Process of infection, *p.* 122. (4) Interference, *p.* 126. (5) Incompleteness, *p.* 129. (6) The dynamics of influenza virus multiplication in the allantoic cavity, *p.* 138. (7) Recombination phenomena, *p.* 140. (8) Mutation, *p.* 151. (9) Summary, *p.* 155.

V The Scope of Biological Generalization 158

(1) Information theory in biology, *p.* 163. (2) The application of pattern concepts to biological problems, *p.* 171.

References 180

PREFACE

A number of more or less fortuitous circumstances have led to the production of this speculative essay on protein synthesis.

In the first place the monograph ***Production of Antibodies*** (Burnet and Fenner, 1949) was due for revision. The theoretical approach adopted in the monograph appears to have provoked considerable interest and to have helped to initiate some fruitful experimental work. It seemed desirable, therefore, to attempt to bring the account up to date.

The second stimulus was our current intense interest in the genetics of influenza virus and the findings by Ada and Perry of the unique importance of ribonucleic acid (RNA) in its structure. Influenza virus and the susceptible cell represents almost the only system in which chemical and genetic aspects of replication can be conveniently studied together.

In the 1949 discussion of antibody production much use was made of the analogy with the formation of adaptive enzymes. The great advances recently made in the understanding of enzyme synthesis in micro-organisms provided a third reason for broadening the scope of any new discussion of antibody production.

Finally, when a first draft had been nearly completed Green's paper appeared in which the self-marker concept of Burnet and Fenner was tentatively applied to the phenomena of carcinogenesis and tumour transplantation. This stimulated a further extension.

If the essay has any virtue, it may be in stimulating workers in one or other of four very different fields to appreciate how developments in all four are converging towards a common point of view.

PREFACE

I am indebted to Dr A. Gottschalk for help in the discussion of enzyme action and to my biological colleagues for reading other parts of the manuscript.

F. M. BURNET

MELBOURNE
1 September 1955

CHAPTER I

INTRODUCTION: ENZYME ACTION AND PROTEIN SYNTHESIS

THE ESSENCE OF LIFE is the replication of specific pattern.

We are concerned with an attempt to understand the significance of this, to point out the difficulties of considering, even at a purely theoretical level, the application of the standard physico-chemical approach to biological matters at this level and to try to develop a series of concepts in terms of macromolecular pattern which may make such matters more amenable to an effective scientific approach.

No one could have the slightest hope of producing a lasting achievement from such an attempt. It seems to be of the nature of the relation between the human mind and the events which make up the universe that the approach to control and understanding is a process in which success leads always to the envisaging of more problems than it solves. At every stage in the past and at every stage in the future, the advancing edge of knowledge in every field has been and will be in a state of confusion. There are phases when the emergence of a new technique or, more rarely, of a fertile generalization allows a swift development of a new area in which ignorance and confusion can be replaced by understanding and the possibility of control and utilization for the satisfaction of human desires. But the edge where ignorance lies beyond the zone of *ad hoc* hypothesis and inadequate experimental technique is always there. Speculation and tentative generalization, as well as the search for and development of new technical approaches, are the legitimate weapons to take us further toward the always receding periphery.

We shall be concerned almost wholly with the properties of protein and nucleic acid simply because these are the types of biological material which (i) seem to be of central importance to the problem, (ii) show evidence of specific function beyond other types of material, and (iii) have been susceptible to a wide variety of experimental approaches at both biological and chemical levels. Possibly the most important reason of all is that in the field of work that has interested me personally, immunology and virology, there are many striking examples of specificity in which the need for some concept of macromolecular pattern seems specially urgent. We cannot for a moment forget about the importance of other components of living organisms beyond functional protein and nucleic acid. Morphology depends on the laying-down, in appropriate fashion, of a host of structural materials —proteins like keratin and collagen, cellulose and chitin amongst the polysaccharides, with many types of inorganic reinforcements, silica or carbonates and phosphates of calcium.

There is, too, a relatively strict control of the inorganic ions, which in animals show their characteristically different distribution in intra- and extra-cellular environments. And even at the more conventionally functional level we have the enormous array of lipids in organisms. In higher animals we find fats that appear to be a relatively simple means for the storage of fuel but, in addition, a wide range of phospholipids clearly of more importance than has yet been ascribed to them, and sterols of many types, some of them hormones of high importance and great subtlety of action. Polysaccharides, mucoids and mucoproteins, such as the lipids, fulfil a wide range of functions in the animal body, as structural and lubricating components, as stores of fuel (glycogen, for instance) and as highly specific patterns conferring serological character on cells and hormonal character on agents such as gonadotrophins. No less than the pattern of an enzyme or an antigen, the distribution and functioning of all these agents is implicit in the patterns carried by the fertilized ovum. Their biosynthesis under gene control is just as important a series of problems as those with which we shall be concerned.

If a start has to be made, however, it must be with the proteins and nucleic acids. They are the constituents *universally* present in living material. The smallest and simplest viruses have no constituents other than protein and nucleic acid. All enzymes are protein with or without prosthetic groups or coenzymes of other nature. And wherever protein is synthesized in an organism we find nucleic acid present. There are striking functional differences between the two classical types of nucleic acids, those containing deoxyribose (DNA) and those with ribose as the sugar component (RNA). In DNA the purine bases are adenine and guanine and the pyrimidines, thymine and cytosine, in RNA thymine is replaced by uracil. There are hints that perhaps small amounts of other nucleic acids may exist, derived largely from the unusual composition of the DNA of bacteriophage T_2 where the cytosine is replaced by 5-hydroxymethylcytosine. It is, however, still orthodox to keep the two types separate and to ascribe to DNA the essential function of carrying the genetic features of all higher organisms, and to look to RNA for some function intimately related to the synthesis of protein.

The characteristic patterns with which we are concerned in experimental biology are those which confer specificity on functional proteins, enzymes, hormones, antibodies and antigens. The central feature of this or any other discussion of macromolecular pattern must inevitably be the nature of the specificity of such proteins and the ways by which the patterns concerned are synthesized or replicated.

It may be that, in the nucleus and in the course of replication of bacterial viruses, protein synthesis is directly controlled by DNA. Elsewhere it seems highly probable that RNA is in some way the controlling agent that confers specificity on the protein being synthesized in the cell. Any discussion of the material basis of life—the means by which replication of pattern is possible—must today be centred on the behaviour of proteins and ribonucleic acids.

Many biologists would probably accept the optimistic point of view that the further understanding of biological processes, including those which we include as dependent on specific pattern, is merely a matter for the continued application of the currently successful methods of physical and of chemical study. It is obviously necessary at the present time to use cruder concepts such as those of immunology or genetics, but eventually these should be expressible in physico-chemical terms. Anyone who claims that standard methods have nearly reached the limit of their effectiveness must first attempt to indicate clearly those aspects of living chemistry which are not accessible now, or will not eventually be accessible to the standard methods of chemical study. Biochemists can point to a continuing series of successes in the isolation, analysis and often the synthesis of substances of biological significance. The synthesis of the polypeptide hormone oxytocin by du Vigneaud and colleagues (1953) is the latest major achievement, one which might well be regarded as a prelude to eventual success in defining the structure of functional proteins. Oxytocin, however, contains only eight amino acid residues and its synthesis presented an extremely difficult problem to the chemists. The smallest 'standard' proteins with a molecular weight around 17,000 contain about 150 amino acid residues. Some exceptional proteins such as insulin are smaller, with a molecular weight about 6000; the great majority are, however, larger and proportionally complex. Chemical methods are, from their nature, only applicable to pure compounds, i.e. to molecular species which can be collected into a large uniform population. In the case of any large biological molecule, the various fragments into which it can be split must also be sorted out into pure substances and characterized. Then follows the effort to reassemble the parts, to synthesize material of the same structure and with the functional activity of the original. Without ever being able to state the precise point at which technique must break down, we can yet be quite certain that no conceivable development of organic chemistry will provide us with the detailed structure of trypsin or of the particular nucleic acid that can transfer a new antigenic quality from one pneumococcus to another.

There is another feature of the work on oxytocin that calls for comment. The synthetic material, like the natural hormone, provokes contraction of the smooth muscle of the uterus, but there is nothing as yet to indicate what part of its chemical structure is primarily responsible for that action. Even more remote is any knowledge of how the target substance, whatever it may be, of the smooth muscle is related to the oxytocin structure.

Perhaps it may underline our ignorance to recall that the most poisonous protein known, botulinus toxin, appears to be a simple protein with no other components than the amino acids common to all our protein foodstuffs. Someone once pointed out that botulinus toxin contained all the amino acids necessary for the growth of the young rat! Not the slightest clue has been published as to any correlation of its chemical structure with its toxicity. The suggestion by Payling Wright (1955) that botulinum toxin may act as an enzyme perhaps on choline-acetylase at cholinergic end organs is based only on pharmacological evidence.

1. *Enzyme specificity*

If proteins were 'chosen' as the material basis of living matter for one reason more than another it may well have been for their potential versatility as specific catalysts. Next to the formulation of the process of replication, an adequate generalization of the basis of enzyme action is the greatest prize for the academic biologist of the future.

There are various grades of specificity in enzymes even when we confine ourselves to enzymes catalysing well-defined reactions involving relatively small molecules as substrates. Some, notably lipases, appear to be specific only in regard to the type of linkage that is split. Others, of which β-glucosidases may be taken as examples, act only if the bond to be broken and the chemical pattern on one side of the bond are of a certain nature but are indifferent to the nature of the rest of the molecule. Still others appear to be specific for one substrate alone.

An enzyme is a functional concept and it may be that different enzymes have very little that is common at the level of chemical structure. Although it is easy enough to handle the extracellular enzymes of the digestive tract under biologically normal conditions, it is virtually impossible to provide an environment in which to test the function of intracellular enzymes that has any resemblance to the natural intracellular milieu. It is always found that there is a certain optimal range of pH for the activity of a given enzyme and very often there are other ionic requirements as well, a certain level of Ca^{++} ions, for instance.

At a more complex level it may be found that substances other than proteins are needed to allow the activity (or full activity) of enzyme on substrate. Sometimes diffusible substances of relatively small molecular size must be associated with protein to allow the system to function as an enzyme. The otherwise inert protein is then referred to as an apoenzyme, the diffusible component of the system as a coenzyme. There are other enzymically active complexes in which a non-protein prosthetic group is rather loosely combined with protein so that, by appropriate manipulations, it can be removed or manipulated chemically without destruction of the protein position of the complex (apo-enzyme). Finally, there are many enzymes, probably a majority, which are simple proteins built up solely of amino acid units.

An enzyme is conventionally named according to the nature of the substrate and of the chemical change wrought on the substrate. But the 'biological' character of enzyme action becomes apparent when an attempt is made to say whether two enzymes from different organisms are the same or not. By testing them on the same substrate and following the course of purification by their activity on that substrate as a specific criterion, one may reach eventually two electrophoretically homogeneous protein solutions. These act on the same substrate and may, therefore, be called isodynamic. Detailed study, however, will almost always show (i) that the activity per microgram of one is greater than the other, (ii) that, if a range of different substrates is available for comparative study of the two enzymes, quantitative or perhaps absolute differences in the susceptibility of one or more of these substrates will be found.

It will, therefore, usually be impossible to define an enzyme in terms of its complete range of catalytic activity—in most instances it will be impractical to test more than a small fraction of the possible substrates. In practice, enzymes are recognized to be present as a result of tests on selected substrates that can be conveniently studied. The fact that enzymes of different provenance differ in the details of their action is simply something to be accepted as of the nature of things. Equally, we may soon have to recognize that many preparations conventionally regarded as of a single enzyme can be shown by refined methods to contain a mixture of related but not identical enzymes. These anomalies are based presumably on the fact that functionally similar proteins need not be built up of the same sequences of amino acids.

Insulins from different mammals are physically and physiologically similar, but their amino acid sequences differ. So every enzyme is, in whole or in part, a protein molecule and its access to substrate and its activity are probably both influenced by the nature of the groups adjacent to the enzymically effective groupings.

The mechanism of enzyme action may well differ in different cases, but there seems to be a sufficient concordance of opinion to allow some general statements. The most important is that enzyme action is initiated by union between enzyme and substrate. In a few instances it has been possible to provide a direct demonstration of such intermediate compounds. Since the enzyme molecule is in general very much larger than the substrate molecule, the combining groups of the enzyme represent only a small active patch on the surface of the molecule. Following Gottschalk (1953), we may ascribe to the co-enzyme, prosthetic group or active patch the function of providing the active grouping responsible for the attack on the susceptible group of the substrate molecule. For the rest of the molecule, the apo-enzyme, we can deduce several functions concerned with attachment to substrate and activation of the substrate. In general, we must assume a three-dimensional orientation of the substrate-attracting groups on the enzyme surface to allow a complementary 'fit' with substrate that is close enough to effect chemical union or allow electron or proton transfer to take place. In this fitting of complementary patterns, van der Waals force, hydrogen-bond force, dipolar and ionic forces may all play their part. Pauling (1948) has suggested that substrate is 'chemisorbed' to enzyme in such a way as to assume a strained configuration resembling the activated complex for the catalysed reaction. This conception of a complementary relationship which, however, is not a perfect fit and produces a potentially unstable complex, is one which seems to play a particularly important role in the understanding of the interactions of biologically significant macromolecules.

In a slightly paraphrased form we may quote Gottschalk's (1953) conclusions in regard to enzyme action as follows. The ability of proteins to form surface profiles with specific patterns allows them to select as specific substrates compounds complementary to the patterns concerned. Charged, polar and hydrogen-bonding groups in the substrate-combining area of the enzyme attract and bind the substrate in a multi-point contact. By this contact between complementary but not completely juxtaposed groups, the enzyme protein with its rigid structure distorts the substrate to a configuration approaching that of the activated state. The whole process is directed towards lowering the energy of activation for the catalysed reaction as compared with the spontaneous one.

It will be the theme of this discussion that enzymes must be taken as the prototypes of functionally specific protein. This has been explicitly or tacitly recognized by many biochemists, as is shown by the great current activity both in classical enzymology and in the study of adaptive enzymes. It is specially significant that much of the work concerned in determining the conditions for protein synthesis use the activity of an adaptive enzyme as the index of synthesis of specific protein.

The production of a wide variety of specifically patterned proteins is a normal function of every living cell. It is characteristic, however, of any biological function that it can only be analysed and understood when ways become available by which it can be experimentally modified in response to a defined stimulus. When we are concerned with specific pattern of functional protein, the approach to understanding will require that by some manipulation we should induce a cell, a tissue or an organism to produce a recognizably new type of protein. In one sense this can be done by supplying glycine or some other amino acid labelled with a radio isotype which will subsequently identify the protein in which it has been incorporated. This approach has many potentialities for the future, but the most relevant current approach is the study of the conditions under which functionally abnormal protein is produced. Three originally quite distinct disciplines have in recent years converged to contribute each in its way to this approach to the problem of protein biosynthesis. These are concerned with (i) adaptive enzyme production in micro-organisms under the stimulus of the appropriate substrate or inducer; (ii) the production of antibody in warm-blooded vertebrates in response to the appropriate administration of antigen; (iii) the production of new virus protein from the cell infected (stimulated) by pre-existent virus.

The constitution and character of proteins, the universality of enzymes and the basic biochemical resemblance of all living organisms demand that a general mechanism of protein synthesis must exist and that an adequate specification must be able to cover the phenomena of all three fields.

2. *Adaptive enzymes in micro-organisms*

One of the most elementary applications of enzymology is the provisional identification of coliform bacilli by their ability to ferment lactose. This is due to the action of an enzyme β-galactosidase which has become the classical instance of an adaptive enzyme. If a culture of *Escherichia coli* is grown in the absence of lactose or any other β-galactoside and then transferred to a lactose-containing medium, fermentation will commence only after a lag period during which β-galactosidase is accumulating in the cells. If, however, it has been grown in the presence of lactose, the cells contain a high complement of the enzyme and lactose fermentation commences immediately after the cells are brought into contact with the sugar.

E. coli strains which possess β-galactosidase as a constitutive enzyme are not found in nature, but have been produced by rapid alternate cultivation in media containing glucose and lactose respectively as sole carbon source (Cohen-Bazire and Jolit, 1953). Under these conditions a constitutive enzyme will clearly provide a survival advantage to a mutant in which it appears. Monod and Cohn (1952) find that the constitutive enzyme is biochemically and immunologically identical with that induced in the standard strain. It is of interest that this holds also for the β-galactosidases produced by *Shigella sonnei* and *Aerobacter aerogenes*. The similar enzymes produced by species of *Lactobacillus* and *Saccharomyces* are, however, quite distinct immunologically from the β-galactosidase of *E. coli* and from each other. Manson, Pollock and Tridgell (1954) have recently reported similar findings for penicillinase. Pollock's standard strain of *Bacillus cereus* produces a small but measurable amount of penicillinase in the absence of the specific inducer. This is functionally and immunologically identical with the induced enzyme. The same identity of constitutive and adaptive penicillinase was shown for a *B. subtilis* strain. Enzymes from *cereus* and *subtilis* were, however, quite distinct immunologically and showed significant functional differences. It will be necessary, therefore, to assume that, basically, constitutive and adaptive enzymes are produced by the same processes.

The production of the adaptive enzyme β-galactosidase has been closely followed by Monod, Cohn and collaborators. Cohn and Torriani (1953) showed that, from bacteria in which adaptive enzyme production had been induced, an immunologically identifiable protein Gz could be obtained which was absent in cells grown in the absence of an inducer. The normal cells, however, have a physically similar protein Pz which reacts with anti-Gz serum but differs from Gz in three significant respects. Pz is not antigenic in the full sense of being able to provoke the formation of antibody when injected in a rabbit, it has no enzymic action, and it is destroyed by trypsin, while Gz is not. Since only those bacteria which have Pz are capable of responding to produce β-galactosidase, there is evidently some significant relationship between Pz and the synthesis of the enzyme Gz. There does not, however, appear to be any evidence for the conversion of Pz to Gz. Gz is produced exclusively under conditions allowing the synthesis of new protein. The usual stimulus to the production of the enzyme is a substrate such as lactose; but Monod, Cohen-Bazire and Cohn (1951) showed that induction need not necessarily be by a substrate. Melibiose, for instance, is a potent inducer, but is not a substrate for the enzyme. Inducers must have an intact galactosidic radical, but their activity as inducers is quite independent of their affinity for the enzyme.

Monod and Cohn's (1952) view is that the specific inducer combines transitorily or otherwise with some cell component and that it is this complex which provokes the synthesis of β-galactosidase. The complex is a short-lived one as there is a rapid disappearance of the lactose-fermenting capacity on transfer of the culture to medium not containing inducer. The evidence indicates that all the conditions needed for protein synthesis must be provided if the production of adaptive enzyme is to occur. Monod *et al.* (1952) find that all necessary amino acids must be present and that there is a linear relation between growth, as measured by total protein synthesis, and induced synthesis of enzyme.

This galactosidase system of *E. coli* is the most extensively studied example of adaptive enzyme formation, but there are many other examples which have been recognized since Karstrom first pointed out the difference between adaptive and constitutive enzymes. Some of these will need to be mentioned in the discussion. Pollock's (1953) work on the production of penicillinase by *B. cereus* is probably the most interesting. It has still to be discovered, however, whether the almost startlingly effective response of this organism to penicillin has any direct biological significance. It may be that an ability to deal with antibiotics produced by other micro-organisms is needed for the survival of saprophytic bacteria like *B. cereus* in nature. On the other hand, there is also the possibility that penicillin has an essentially accidental action on a mechanism evolved to deal with an unrelated metabolite having some structural features in common with penicillin. The existence of small but definite amounts of similar enzyme in culture fluids from *B. cereus* grown in the complete absence of penicillin points strongly towards the latter conclusion. No indication of the nature of the normal metabolite seems to have been obtained.

In this example only a transient contact with penicillin is needed to produce a persisting capacity to synthesize penicillinase in a penicillin-free medium. Growth of the induced culture in penicillin-free medium gives rise to a linear production of penicillinase despite the logarithmic increase in the total protein with growth of the culture. The inducer, therefore, is apparently not a self-replicating agent nor is any mechanism which it may call into existence capable of replicating itself.

As we have already mentioned, the current concept of protein synthesis suggests a process in which nucleic acids are intimately concerned and the first approach to an analysis of adaptive enzyme formation will be in relation to DNA and then to RNA function.

There is good evidence that DNA has no *immediately* necessary part to play in the formation of adaptive enzymes. The capacity to produce an adaptive enzyme of some specific character in the presence of the appropriate inducer is a genetic character determined presumably by some DNA-carried pattern of the nuclear material. The function of DNA then would be simply to direct the fabrication of machinery that would allow the inducer to function. In support of this view we may cite the recent evidence of Pardee (1954) that β-galactosidase can be produced in cells which have been treated with doses of X-rays or nitrogen mustard sufficient to destroy all turnover of DNA phosphorus. Rather striking evidence in the same direction comes from Cohen and Barner (1954), who showed that a thymine-requiring mutant of *E. coli* could produce adaptive enzyme in the absence of thymine which is the pyrimidine base needed for DNA synthesis but not for RNA synthesis.

A different line of evidence from another system is due to Hotchkiss and Marmur (1954), working with pneumococcal transforming principles. A strain of pneumococcus is known in which a mannitol-utilizing enzyme can be induced in the presence of the specific substrate. This quality can be transferred to a strain that lacks it, using the standard technique with purified DNA from the mannitol-fermenting strains. This succeeds irrespective of whether the donor culture has or has not been induced to produce the enzyme. It is a capacity to develop a mannitol-utilizing mechanism, not a capacity to utilize mannitol that is transferred by the transforming principle.

There is equally strong evidence that RNA on the other hand plays a very important and direct part in the process. Pardee (1954) showed that *E. coli* mutants which need purines, pyrimidines or phosphate for growth also require these for induction of β-galactosidase. Within a few minutes of the exhaustion of the critical component, production of adaptive enzyme ceases, an effect which can be shown not to be due to blocking of amino acid synthesis or to destruction of existing enzyme. It is noteworthy that the relatively large amount of RNA pre-existent in the bacteria is apparently unable to provide building blocks for whatever is needed for adaptive enzyme production. This is in line with Manson's (1953) finding that there is little or no turnover of C^{14} in bacterial RNA. It seems that once RNA has been synthesized it is available only for the specific function for which it was synthesized. It cannot, in *E. coli* at least, provide a pool of purines and pyrimidines from which new patterns of RNA can be fabricated.

Gale and Folkes (1954) used a different system to reach very similar results. By sonic disruption of staphylococci and extraction of most of the nucleic acid with 1 M NaCl, they obtained a system that could be used for a wide range of metabolic experiments. RNA can be synthesized by this system if appropriate pools of purines and pyrimidines and of amino acids and glucose are provided. The rate of synthesis of RNA can be conveniently followed if C^{14}-labelled uracil is included in the mixture. When β-galactosidase is stimulated into production by the addition of an inducer, there is a large increase in the incorporation of uracil. In this particular system it appears that DNA must be present for RNA synthesis since this fails to occur if the preparation is treated with deoxyribonuclease, but this does not necessarily mean that DNA must undergo resynthesis. RNA is essential for the synthesis of adaptive enzyme as shown by failure of the latter to occur in the presence of ribonuclease. Further, indirect evidence in the same direction is given by the fact that β-galactosidase production is inhibited by penicillin which at the same time prevents the synthesis of the presumably specific RNA.

In *B. cereus*, Pollock (1953) found that the initial stages (the first 14 minutes) of penicillinase production were sensitive to ultra-violet light, suggesting that RNA was intimately concerned at this stage. With *E. coli*, Swenson and

Giese (1950) found that the wave-lengths of ultra-violet light most effective in inhibiting enzyme production were those preferentially absorbed by nucleic acids.

These results from the study of adaptive enzyme synthesis in micro-organisms can perhaps be summarized and in part interpreted as follows :

(1) Final genetic control is vested in the DNA which in some way is responsible for providing the reaction mechanism.

(2) The inducer acts by combining with some cell component to form an organizer which develops or activates a mechanism in which RNA is an essential component and which then goes on producing the new protein at a linear rate.

(3) For the production of this mechanism all the necessary components for the synthesis of RNA must be present, and under the influence of ultra-violet light of wave-length 2650 Å. the production fails to occur.

(4) There is no evidence that this mechanism is a self-replicating one.

(5) For new protein synthesis a pool of all necessary amino acids must be present.

Evidence has already been given that, in the two adequately studied instances of β-galactosidase and penicillinase constitutive and adaptive enzymes from the same species of bacteria are functionally and immunologically identical. As is well known, the borderline between adaptive and constitutive enzymes is by no means sharp and introduction of substrate into the environment of a bacterium capable of producing a constitutive enzyme will usually result in a marked increase in the production beyond that obtained in a medium lacking the substrate. It may be sometimes difficult, however, to decide whether the observed increase in the amount of enzyme is actually due to new enzyme production or to release or activation of pre-existent but inactive enzyme.

We shall take the point of view that the differences are not in the enzyme as such but in the persistence and replicability of the mechanisms by which they are synthesized.

3. *Chemical aspects of the biosynthesis of protein*

Early thought on the synthesis of proteins in the cell tended to centre round the idea that they might be built up from amino acids by the reverse action of proteolytic enzymes. The insuperable difficulty from this point of view was the existence of many different types of protein and the necessary postulation of ever larger numbers of types of enzymes to build up these specific patterns. Each enzyme, being itself a protein, would demand in its turn a battery of enzyme-synthesizing enzymes, and so *ad infinitum*. The only reasonable solution was to look for some template mechanism by which a replica can be built up of some pre-existent pattern. It is probably true to say that in one form or another this idea dominates most recent thinking on the subject.

Haldane (1954), for instance, points out, primarily from a geneticist's point of view, that the cell must have means by which a large number of models can be copied with great fidelity. In his view the current hypotheses fall into two categories.

(1) The model is in the form of an extended one- or two-dimensional structure on which components from a pool of 'building blocks' will lay down a similar structure by a process analogous to crystallization. Haurowitz (1952*a*) has developed a theory of protein synthesis along these lines.

(2) The specific pattern of the model is copied (or re-coded) in a completely different medium, giving a negative or template from which two (or more) new positive copies can be produced. As far as I am aware, all theories of this type postulate that protein and nucleic acid are the media that carry the two equivalent but complementary patterns.

The chemical aspects of proteins which have been regarded as particularly relevant to the problem of their biosynthesis are (i) the nature of the primary polypeptide chain and (ii) the way in which it is folded to give a globular protein.

It is not clear whether branched polypeptide chains exist, but there are certainly protein molecules which have more than two terminal amino groups. Hæmoglobin, for instance, has six terminal valine groups. Everything indicates that there are at least long sections of unbranched polypeptide in which the sequence of amino acids is responsible for the individuality of the molecule. It is still, however, uncertain whether the configuration of the functioning protein molecule (an enzyme, for instance) is determined simply by the sequence of amino acids in the chain or chains involved and the non-specific ionic or other conditions under which the coiling or folding of the chain takes place.

Haurowitz (1952*a*), influenced very largely by immunological considerations, believes that, for most proteins at least, the formation of the polypeptide chain is only a preliminary step in biosynthesis. The specific functional character as enzyme, antigen, hormone or antibody in his view is conferred by the subsequent folding of the primary polypeptide chain in direct contact with an appropriate secondary template. Some evidence bearing on this concept can be obtained from the behaviour of proteins at an air-water interface where most soluble proteins spread out into virtually two-dimensional sheets. There are many examples where such spreading involves denaturation with loss of functional specificity and to this extent Haurowitz's concept is supported. More important, however, is the fact that in the case of some enzymes the expanded protein can be recovered in functionally active form. Kaplan (1955), for instance, draws attention to the fact that catalase 'expanded' at an air-water interface has properties almost the same as those of catalase as it exists intracellularly and quite different from those of the extracted enzyme. This points strongly against the necessity for a secondary template.

Perhaps largely on account of the greater simplicity of the concept, there seems to be a general trend to assume that the specificity of a protein is conferred at its primary synthesis. If this is so, the problem of synthesis in its broadest terms would be to determine how amino acids, *A*, *B*, *C*, etc., were linked together in some predetermined linear sequence, and then liberated from the matrix carrying the code that determined the sequence of amino acid residues. The main possibilities are:

(1) There is a coded sequence on the template in one- or two-dimensional form *a*, *b*, . . . , *x* which allows the placing of the amino acid residues *A*, *B*, . . . , *X* in the correct sequence from an amino acid pool, while some appropriate enzyme process sees to the formation of the peptide links holding them together. This might result in the almost simultaneous completion of the whole chain which is then presumably floated off the template but, as suggested by Dalgliesh (1953), the construction of the peptide chain might be going on only at one point in the chain at any moment, release taking place as soon as each segment is added (Fig. 1). Zipper-like processes of this type tend to appear rather frequently in current hypotheses on the borders of chemistry and biology.

(2) Relatively large segments are fabricated elsewhere and brought together into the final form at the template. This might allow a relatively large number of functional patterns to be made up from a few standard sub-patterns.

Chemical evidence in favour of one or the other of the hypotheses can be sought by following the incorporation of amino acids labelled with radio-isotopes into protein produced at some site of active synthesis. Campbell and Work

(1953) have used this approach to study the production of milk protein in the mammary gland. On the whole, their results are compatible with a mechanism in which the whole chain is built up more or less simultaneously on a template from an amino acid pool. There are some results which may indicate the incorporation of relatively large peptides, but could be equally fitted to an elaboration of Dalgliesh's hypothesis in which the same template may be simultaneously synthesizing more than one polypeptide chain.

Fruton (1950) has shown that proteolytic enzymes can catalyse the exchange of one amino acid for another within a peptide chain—the process of transpeptidation—and there is no thermodynamic objection to supposing that protein synthesis takes place by transpeptidation under the influence of intracellular cathepsins. Against this hypothesis is the absence of free peptides and the difficulty of utilizing peptides when these are supplied ready-made and labelled. Rather direct evidence against it is also provided by Loftfield *et al.*'s (1953) work with the non-physiological amino acid, amino-butyric acid. Artificial peptides containing amino-butyric acid residues are split by rat liver cathepsin, but when the labelled amino acid is injected it is *not* incorporated into liver protein *in vivo* although it can be shown to penetrate the cells and presumably enter the amino acid pool.

Spiegelman and Halvorson (1953) found that the synthesis of the adaptive enzyme maltase by yeast involved the active participation of free amino acids. They used amino acid analogues and showed that the presence of any one in effectively blocking concentration prevents the incorporation, not only of its corresponding homologue, but also of all the other amino acids of the pool. They conclude that the first stable intermediate formed in protein synthesis is already of such complexity as to demand the utilization of a large proportion of the various amino acids.

Although few would claim that the evidence is decisive, the present opinion amongst those actively working in the field is clearly in favour of a template hypothesis in which probably in some zipper-like fashion polypeptide chains are fabricated directly from an amino acid pool.

Such a conclusion in itself has no particular relevance to the nature of the template involved.

The part played by ribose nucleic acid. Present-day interest in nucleic acids is largely centred on DNA as the characteristic component of the nuclear mechanism and the probable bearer of genetic characters. The double helix formulation of Watson and Crick (1953) has been very rapidly accepted and Delbruck (1954) has recently propounded a mechanism by which two replicas can be reasonably produced from a pool of the necessary building blocks. Here at least the trend is strong towards a mechanism of direct replication in which the sequences of nucleotides with paired bases, adenine-thymine (A-T), guanine-cytosine (G-C), provide a direct guide to the sequence being laid down in the new chain.

The Watson-Crick formulation, however, appears to provide no significant clues as to the relation by which the presumed binary code—in which, as it were, A-T and G-C units take the place of the dots and dashes of the morse code—is translated into the chain of action leading eventually to the synthesis of specific protein and other cell components.

There are some faintly unsatisfactory features about the standard assumption that this is the basic code of genetic structure. There is, for instance, much to suggest, as Haldane (1954) points out, that at certain phases of meiosis the chromosomes may be almost wholly protein. Tomlin and Callan (1951) found that chromosomes from amphibian oocytes at the diplotene stage are protein threads with little or no nucleic acid. Under the electron microscope they are about

200 Å. in diameter and show no signs of doubleness. Writing before the Watson-Crick formula appeared, Haldane suggested that this thread might act as a template on either side of which a polynucleotide chain was laid down. A simple basis for the phenomenon of crossing over could be provided on this hypothesis.

Another difficulty is in regard to the amount of DNA present in a mammalian cell nucleus or in a bacteriophage head. A molecule of DNA with molecular weight of 1 million would contain about 3000 nucleotide residues, i.e. about 1500 'dots and dashes' which could be expected to carry information or control corresponding to many genes. A mammalian cell presumed to contain 20,000–40,000 genes could provide a complete DNA molecule of molecular weight 1,000,000 for each gene and still utilize only 0·5–1 % of the DNA in the nucleus. There seems to have been little discussion in regard to the function of the 'unnecessary' DNA.

Much less is known about the physical and chemical nature of RNA. Chemically the units are said by Todd (1954) to to be β-D-ribofuranosides arranged in a form by which the bases are attached to carbon 2 of each ribose unit while the ribose units are joined through phosphate groups linking carbon 3 of one to carbon 5 of the next.

This contrasts with DNA in which the bases are joined to the carbon in the 3-position of the deoxyribose units and makes it theoretically possible that branched chains may exist in RNA. They appear to be impossible with DNA.

The bases concerned are the same as for DNA except that thymine is replaced by the related pyrimidine, uracil. There is an interesting discrepancy between the RNA obtained from plant viruses and that from most other sources. Classical RNA, e.g. from rat liver cells, resembles DNA in having the ratios of adenine/uracil and guanine/cytosine approximately

1 and, therefore, making a formula equivalent to that of Watson and Crick for DNA highly probable. This regularity is absent, however, in plant virus nucleic acids, and there is a large deviation from the rule in RNA from mammalian nuclei according to Elson and Chargoff (1954). There is no real evidence for or against the view that DNA may act as a template for RNA production.

The average size of the RNA unit is not known, but its molecular weight is probably less than 100,000.

It is perhaps highly relevant that ribonucleotides form some of the most important substances in intermediary metabolism. They include adenosine di- and tri-phosphates, co-enzymes I and II and co-enzyme A. The idea that ribonucleic acid is associated with protein synthesis was first suggested in 1937 by Caspersson and has been developed by Brachet and many others since then. Dounce (1952) has provided the most detailed picture of the possible template action of RNA in the light of what is known of the function of the physiologically active nucleotides. He believes that RNA provides a template which is first phosphorylated to give a polyphosphate analogous to ATP with energy-rich phosphate bonds which react with amino acids to give a ⌜nucleic acid, peptide⌝ complex. Linear polymerization eventually gives the complete polypeptide.

If RNA does carry a 'coded' pattern by which the synthesis of a polypeptide chain can be directed to give the 'right' sequence of amino acids, the nature of the process by which the code is, as it were, translated from one medium to another is obviously of the greatest importance and interest. A dozen different kinds of circumstantial evidence point in that direction, but no decisive proof of the association or convincing picture of the mechanism involved has yet been given. Most of the indirect evidence will be discussed in relation to the three main topics of this work, adaptive enzymes in micro-organisms, antibody production and virus replication. Briefly, we have:

(1) the intimate association of RNA synthesis, as measured by the incorporation of labelled building blocks, with the production of new types of protein, especially adaptive enzymes;

(2) histological evidence of accumulation of RNA in cells where active protein synthesis is occurring, either by Caspersson's methods with ultra-violet light of known wavelength or by special staining methods, notably with pyronin;

(3) blocking of protein synthesis in appropriate preparations of cellular fractions by the action of ribonuclease (Allfrey *et al.* 1953; Gale and Folkes, 1954); and

(4) the presence of RNA as the only component other than protein in the macromolecular plant viruses.

On the basis of these findings we shall use the hypothesis that RNA serves as a template for protein synthesis and try to follow the implications of this hypothesis in the three fields of adaptive enzymes, antibodies and viruses.

General theories of protein synthesis. We shall adopt many features from the concepts put forward by Haurowitz, Brachet, Dounce, Pollock and others and assume that protein is synthesized as a polypeptide chain in expanded state and possibly under stress in contact with RNA. The RNA has three functions.

(1) It holds the protein as an extended polypeptide chain at least over the segments in which the final stages of synthesis are taking place. This semi-activated condition may be of great importance from the point of view of enzyme synthesis.

(2) It provides energy-rich phosphate bonds which will provide the necessary energy for transpeptidation or other reactions needed in the synthesis. Dounce (1952) is primarily responsible for this concept.

(3) It carries the code which determines the sequence in which amino acid residues are added. This is a vital feature and one that is extremely difficult to visualize. Gamow and Metropolis (1954) point out that with 20 amino acids the number of possible sequences is very much larger than could be coded by the sequence of 4 bases (or 2 purine-pyrimidine pairs) in a polynucleotide chain.

It is quite impossible at the present time to picture a nucleic acid template which, in the form of a single chain or Watson-Crick helix, could provide a specific linear sequence for polypeptide synthesis, each amino acid added corresponding to a structural unit in the polynucleotide chain. There must be some subsidiary mechanism or, in Gamow's words, 'the code must be extremely restrictive and must lead to strong intersymbol correlation between the neighbouring amino acids in known protein sequences'. At the present time it appears necessary to reserve judgment and to leave open the possibility of some associated factor, histone possibly, which allows a sufficient complexity to carry the needed coding.

The recent work of Gale and Folkes (1955) may open up an important direct approach to the type of correspondence between nucleic acid structure and the sequence of amino acids synthesized. Working with disrupted staphylococci, which in the absence of RNA failed to incorporate labelled amino acids, they found that this capacity could be recovered if staphylococcal RNA were added and also if the same RNA broken down to di- and tri-nucleotides by ribonuclease were used. Using labelled aspartic acid, they found that its incorporation could be ensured by a simple di-ribonucleotide (adenine-cytosine), but that this was ineffective with leucine-glycine or glutamic acid.

In a later section the suggestion will be found that an easily separable relationship between RNA and protein need not necessarily take the form of a simple linear apposition and that in some more complex arrangement a more elaborate code could be carried by the polynucleotide without necessarily assuming any additional effect due to other factors. The evidence for this is derived solely from virus studies and its discussion is best postponed until virus replication is considered.

In all subsequent discussion it will be convenient to use the term RNA+ for the template of the protein-synthesizing mechanism and when 'RNA+ building blocks' or similar term is used it will cover any components needed for the synthesis of the whole structure. The physical relation between the polypeptide chain being synthesized and the RNA+ template is quite unknown. Primarily, to allow simplicity in diagrams, it will be assumed that the RNA code is linearly arranged and that there is a linear correspondence of the sequence of amino acids in the polypeptide chain being synthesized and the sequence of 'built-in symbols' on the RNA+. In all diagrams RNA+ is shown as a black bar, protein as a white bar and where necessary specific features are indicated at corresponding points on each. The main reason for making RNA the chief feature of the protein-synthesizing mechanisms is because the availability of P^{32} makes it possible to gain an estimate of what synthesis of RNA phosphate is going on at significant times. It is an experimental fact that the turnover of RNA is greatly influenced by the state of protein synthesis in the cell concerned.

In bacteria, where RNA is present in relatively large amount, isotopic studies indicate that preformed purines and pyrimidines are accepted and preferentially used for synthesis of RNA (Bolton, 1954). It appears that bacterial RNA once formed is not available for breakdown and reorganization. Manson (1953) found no turnover of C in bacterial RNA, and in Pardee's experiments with purine-requiring mutants of

E. coli adaptive enzyme synthesis became impossible as soon as the supply of the necessary purine was exhausted.

In vertebrate cells the turnover of P^{32} in RNA depends on the type of protein synthesis going on. De Deken-Grenson (1953) studied the pancreas of the mouse in which there is a rapid synthesis and secretion of protein, especially when stimulated. He found no significant difference in the rate of renewal of RNA in quiescent or pilocarpine-stimulated glands. In both the turnover was low, between 0·1 and 0·2% per hour. Hokin and Hokin (1954) obtained similar results and contrasted them with their earlier work on growing feather buds, where they found that the speed of protein synthesis similar to that of RNA synthesis. In the rat liver there is an active turnover of P which Boulanger and Montreuil (1952) find is more active for some nucleotides than others, uridylic acid being the most active and guanidylic the least.

The general impression received is that the new RNA synthesis is not required for routine synthesis of protein as in the secreting cells of the pancreas. It *is* required for growth or for any *change* in the nature of the protein being produced by the cell.

It must be recognized that all discussion of protein synthesis must at present be on a basis of speculation at a functional level. The morphological concomitants are left unconsidered, but we can be quite certain that they will eventually become highly relevant. Much RNA is associated with mitochondria, more with the microsomal fraction of the cytoplasm. Electron microscope studies show the presence of complexly structured mitochondria and very complex, finely sacculated, structures in protein-secreting cells like those of the mouse pancreas. Enzymological studies show intense concentration of enzymes in mitochondria which are obviously highly organized centres of synthetic activity. The same holds, but probably to a lesser degree, for the microsomal fraction.

Sooner or later it will probably become necessary to bring these morphological and enzymological aspects into the picture of protein synthesis. It is clearly essential that speculation should be kept at a general enough level to allow its

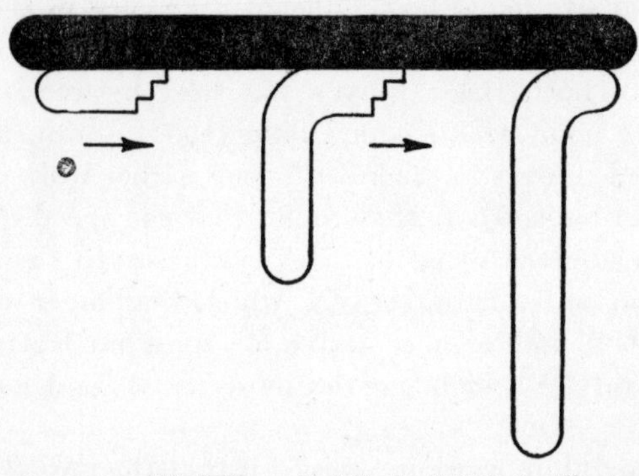

Fig. 1. A figure modified from Dalgliesh (1953) to illustrate how a single template (black bar) might be involved in the synthesis of three (or more) polypeptide chains.

***Note.* In this and similar figures the convention is adopted of indicating protein (polypeptide chain) as a white bar and the postulated template based on RNA as a black bar.**

modification to fit facts in this category as they become available.

We must underline that the mechanism RNA+ may eventually have to be equated to something much more complex than the representation of it that we shall adopt. The essential feature of the hypothesis is that RNA+ has a complementary relationship to the pattern sequence of the polypeptide chain for which it serves as template. Further, the hypothesis of mutual template relationship between RNA+ and protein patterns demands that we assume that some segments at least of the polypeptide chain can, under appropriate circumstances, determine the pattern in which newly synthesized RNA + is laid down. The general character of the approach is indicated by the highly schematized diagrams in Figs. 1 and 2.

It may be worth while first to make some remarks about the general process of replication by the formation of a complementary pattern. There are, of course, many analogies from the arts and handicrafts that deal in patterns. In fact, it is much more difficult to find examples in which a pattern

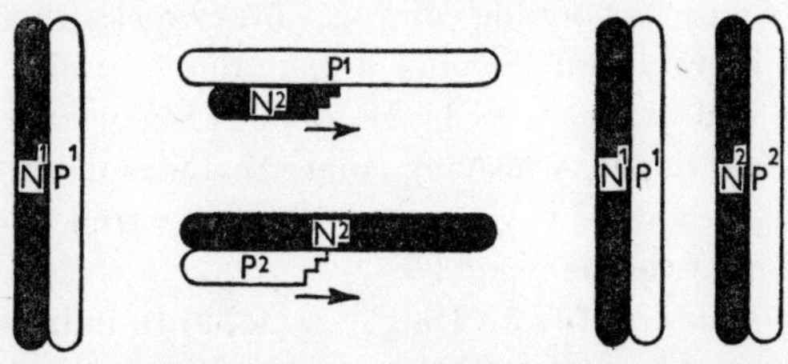

FIG. 2. Scheme to illustrate how a biological unit (NP) might replicate by two successive productions of complementary patterns. Functional separations between the two components must be postulated where necessary. N, nucleic acid; P, protein.

is copied directly by any mechanical process than to enumerate those using a complementary mechanism. In photography we transfer the primary pattern to a negative from which unlimited numbers of positive patterns can be made. For the mass production of three-dimensional objects the standard approach is to produce the complementary three-dimensional pattern, the mould or matrix, and to prepare positive casts from this. Another analogy that may be worth considering is the arrangement of steel punches that produces the pattern of holes in a punch card. This is normally a unidirectional relationship, but there are probably machines in existence in which the arrangement of punches to produce a given pattern is determined by the insertion of a card carrying that pattern. In all mechanical replication processes—and perhaps in all biological ones—a certain distortion of pattern can be expected at each replication. If we use a method by which duplication occurs directly and each product acts as the model for the next duplication, the production of 64 replicates will give for each final product six opportunities for distortion to occur. On the other hand, a method by which the original pattern is used to produce a complementary negative or mould can allow 64 positives to be made from this, each of which will only have had two opportunities for distortion. The double complementary method of replication is a more versatile concept than direct replication and one may well expect it to be adopted in biological material. Nevertheless one may also remember that, despite the virtues of the wheel, nature never succeeded in devising a biological application of that concept!

Fig. 1 is modified from Dalgliesh (1953) to indicate a possible way in which the RNA+ template may give rise to a series of protein molecules, all bearing the complementary pattern and hence duplicates of one another. The essence of Dalgliesh's idea—an application of the 'zipper' family of concepts—is that the production of complementary pattern takes place at only a limited zone and with the completion of the process activity moves forward much in the fashion of the runner of a zipper. It is assumed that the process takes place in the presence of a pool of amino acids and/or peptide sub-units from which the nature of the template can draw the unit needed next in sequence. The hypothesis was introduced primarily in relation to the problem of showing experimentally that a template mechanism was necessarily involved in protein synthesis. At one stage it was claimed that any template theory would demand that the whole pattern be laid down simultaneously in an immediate transfer from a pool of amino acids to the polypeptide chain. Under these circumstances, if labelled glycine is added at a given moment to the pool, the protein produced at intervals thereafter can be examined for the uniformity with which labelled glycine residues are distributed through the length of the polypeptide chain. Under the original view the distribution should be uniform although the proportion of labelled glycine will fall with time. Under Dalgliesh's scheme, as well as under hypotheses in which peptides or incomplete proteins are built and stored for later fabrication as proteins, there is a possibility that different proportions of the labelled amino acid will be present in different parts of the chain, depending essentially on the time when material for the synthesis of the parts in question was withdrawn from the amino acid pool.

The implications of the hypothesis on which Fig. 1 is based have already been mentioned in part. They are (i) that the polypeptide chain in the regions at which it is held to the RNA+ template is under constraint since on release it will fold to take on the globular form, (ii) that the subsequent folding of the peptide chain is determined in all essentials by the sequence of amino acids and is not subject to the specific action of a secondary template in Haurowitz's sense, (iii) that the occurrence of synthesis will depend primarily on the availability of all the necessary building stones at the site. In the absence of any essential unit, the process will stop at once.

The further development of this point of view is more conveniently deferred until the main types of experimental evidence bearing on the modification of protein pattern have been discussed. The nature of enzyme action is highly relevant and the light thrown on the general problems of protein synthesis by the study of adaptive enzymes is the main source from which the general concept can be elaborated.

4. *The nature of adaptive enzyme synthesis*

We can, superficially at least, divide proteins into two groups: structural and storage proteins like keratin, silk, collagen and egg albumin, and functional proteins like enzymes, antibodies and hormones. It is reasonable to believe that the functional ones are the more primitive and the more important, therefore better fitted for discussion in the present context.

Any functional protein or, for the matter of that, any functioning macromolecule of the living cells is derived on the one hand from a producing mechanism and on the other hand acts by specific adsorption with its target substance, typically a substrate. At least three patterns are here involved, and it would be in line with every remotely analogous procedure in human technology to believe that, in many instances at least, the producing mechanism has its pattern 'blue printed' in one way or another by the pattern of the eventual target substance. But one must not be too anthropomorphic or forget that accidental configurations favoured by selective survival must arise at the biochemical level just as at the higher levels of morphology or behaviour.

The work on adaptive enzyme formation by micro-organisms discussed previously provides almost overwhelming evidence that there are no significant differences functionally or serologically between corresponding adaptive and constitutive enzymes. It, therefore, becomes virtually necessary to assume the same essential process of synthesis in each case, the difference between the two being ascribed chiefly to the stability or persistence of the mechanism concerned.

Just as in the higher vertebrates, learned behaviour on a simple background of instinctive trends has replaced purely instinctive behaviour with a minimum of learning, so one might guess that the laying-down of the biochemical patterns of the organism in the sequence of individual development includes a whole variety of interactions by which the dynamic chemical machine is brought into being through successive functional changes until it is an effective organism. The appearance of a substrate, for instance, may act as the stimulus to the appearance of an enzyme to deal with it. Immunological phenomena to be discussed in a later section point strongly in this direction and give more than a hint that a process which, under one set of circumstances, would provoke the temporary production of an adaptive enzyme will, under another set, induce the continuing synthesis of a constitutive enzyme.

If this is a legitimate point of view, we have to picture a means by which substrate can modify protein and protein can modify the mechanism which produces it. The first clue is to be found in the existence of transpeptidation. It is known that proteolytic enzymes can catalyse the exchange of one amino acid for another in the middle of a peptide chain and there is no thermodynamic objection to the possibility of such reactions occurring frequently in the living environment. One might picture then a situation in which a target substance, or more strictly its specific pattern, might, by contact with a developing or newly synthesized segment of a protein, induce by transpeptidation the insertion of amino-acid residues which would allow a new protein configuration complementary to the pattern of the inducer.

It seems possible that a development of Pauling's (1948) theory of enzyme action may lead to a generalization that when a polypeptide chain is held under strain, as in the process of its synthesis, contact with a variety of organic patterns may act as an 'enzyme' for the reshaping of the chain to a complementary pattern. When the constraint is released and the protein takes its natural shape as a soluble protein, it becomes conceivable that the now 'not quite complementary' relationship to the substrate is the basic reason for its effectiveness as an enzyme on that particular substrate. This is virtually Pollock's hypothesis, although he leaves the mechanism open by which substrate (or equivalent inducer) can modify the protein previously produced, into specific enzyme. His development of the hypothesis is to envisage a general protein-forming system on which the final enzyme patterns would be imprinted by the inducer molecule. Much more, however, is necessary than the modification of one protein molecule by the inducer. Pollock finds that in *Bacillus cereus* the inducer penicillin is fixed to the bacterial cell in amount corresponding to about 100 molecules per cell for optimal production of penicillinase. Calculations show that unless penicillinase has a turnover number much higher than that of catalase, a large number of penicillinase molecules must be synthesized as a result of the presence of the one organizer. If we confine ourselves for the time being to the induction of penicillinase in *B. cereus* where there is known to be a small production of enzyme in penicillin-free medium, we can consider the following possibilities:

(1) Yudkin's theory (1938) of enzyme adaptation was that the inducible enzyme is in equilibrium with a precursor; addition of the substrate or other inducer upsets the equilibrium which is then restored by further enzyme synthesis. Mandelstam and Yudkin (1952) found that simple mass-action calculations on this basis gave a satisfactory agreement with yeasts adapted to ferment galactose and maltose. Pollock's (1950) findings, however, are clearly incompatible with such a formulation. The inducer must act by combining with something other than the enzyme.

(2) The organizer, produced by the union of some cell component and the inducer, acts as a specific catalyst (in our convention, a template) for the conversion of some precursor or pool of precursors to specific enzyme. This is Pollock and Monod's view, and in its general form, can probably be accepted without reserve. Any attempt to go further than this, however, meets with difficulties.

(*a*) Penicillin appears to be firmly united to the cell induced, presumably by a *Specific Complementary Pattern* (SCP) relationship. The pre-existence of penicillinase production at a low level suggests that the likely site of the SCP is in relation to the constitutive enzyme-synthesizing mechanism. We might assume, therefore, that penicillin is attached to constitutive penicillinase still in contact with RNA+ and, therefore, able to combine with, but not destroy, penicillin. The combination, in one way or another, will be responsible for an increased production of enzyme. The most plausible suggestion is probably that the newly stimulated and stabilized protein is a more effective template for specific RNA+production, and that the newly produced RNA is actually responsible for the increased production of enzyme.

(*b*) If we accept the evidence that standard *Escherichia coli* produces *no* β-galactosidase in the absence of the inducer, this interpretation obviously cannot hold. In view of the desirability that a general picture should be available, we might attempt to deduce from the β-galactosidase results a mechanism that would also fit the penicillinase findings. Since there is by hypothesis no pre-existent SCP in the cell for the inducer, it must make a more or less accidental union with some entity closely related to enzyme synthesis. RNA+ units by hypothesis are capable of union with, and/or synthesis of, any pattern of protein so that almost any organic molecule could be expected to find an opportunity of a specific union at one point or other of some such unit. Under appropriate conditions one can imagine that such a union might induce a specific complementary pattern in protein synthesized in relation to the modified template. The possibility that the organizer is a combination of the inducer and the postulated protein-synthesizing RNA + template must be kept in mind.

(c) A compromise view of some attractiveness is that the inducer as such cannot have any influence on the synthesis of protein but acts primarily as a stimulus to the synthesis of an organizer, itself requiring RNA activity. The organizer would then be a modified transpeptidation unit perhaps with the inducer built in which would act as a secondary shaper of precursor-enzyme, both mechanisms operating at the amino acid utilizing stage. This comes very close to Haurowitz's view, the main difference being that the secondary template is assumed to act coincidentally with the primary one.

Amongst the findings which must be considered in any attempt to decide amongst these or other possibilities is Monod and Cohen-Bazire's (1953) work with the mutant of *E. coli* that produces β-galactosidase as a constitutive enzyme. Here the situation is the opposite to what was postulated above, in 2 (*a*). The addition of inducer *diminishes*, at least temporarily, the production of enzyme. This at any rate points again to the existence in the organism of a SCP *to* the inducer, but we have to postulate that the influence is the reverse of that produced by penicillin under similar situations. Whether such differences are reasonable could only be decided when the point has been examined for a wider range of alternatively adaptive or constitutive enzymes.

The possibility described under 2 (*b*) is formally very close to the suggestion that a precursor is converted by the action of the inducer (or its product, the organizer) into enzyme. In slightly different terms it might be said that in the phase of production of the organizer a new *part* of the protein-synthesizing mechanism is laid down. This is in line with the evidence that more than one gene-controlled step may be demonstrable in the synthesis of a given antigen. Fox (1954), basing his opinion on findings of this type, suggests that protein synthesis occurs in two gene-controlled stages—a preliminary stage involving several steps by which a non-specific precursor is produced and a second stage by which final specificity is conferred on the product. Only the last may be due to template action and it may be that the template itself is synthesized by a two-stage process.

The question of a precursor is still controversial. It is by no means generally accepted that Pz is a precursor in any simple sense to Gz β-galactosidase, and Spiegelmann and Halvorson (1953) conclude that precursor cannot be converted into enzyme although it is still possible that enzyme is produced by interaction of precursor with a full pool of amino acids. The evidence, however, points in their opinion more towards the direct synthesis of enzyme from the amino acid pool.

Other evidence for the existence of a precursor (i.e. precursor 1 according to Pollock's scheme, in which precursor 2 would be exemplified by trypsinogen) has been obtained by Koritz and Chantrenne (1954) who studied the incorporation of labelled glycine into the proteins of reticulocytes. Immature red cells have only a short period in which they retain the capacity to incorporate amino acids and the peak of glycine incorporation comes two days before the peak of RNA production. There are three specific proteins produced by these cells which can be readily estimated—hæmoglobin, carbonic anhydrase and peptidases. The production of these is approximately parallel to RNA production. The authors consider their evidence points to the production of an undifferentiated primary protein by a process not involving RNA. This is then differentiated into a biologically active form by the mediation of 'perhaps stereospecific RNAs'.

It seems clear that we must accept the existence of polypeptide precursors of a sort even if we consider them as little more than a convenient way of storing material for the 'amino acid pool'. We can, in fact, see no reason why the production of a complex functional protein may not involve two or more remouldings of polypeptide material first put together in more than one site. All such interactions, however, must be assumed to take place in the presence of an amino acid pool with the mechanism of transpeptidation available, and in relation to a template which will be presumably what we have called RNA+.

Following Pollock (1953) we must somehow postulate a means by which a target substance (substrate) can influence the structure of a protein during the course of its synthesis. From various sources one draws the suggestion that a system in which a polypeptide is being synthesized from an amino acid pool might be susceptible to interference by certain substrate groupings. We know that the sequence of amino acids is determined in some indirect fashion by the configuration of RNA+. It is reasonable, therefore, to assume that, when an appropriate small molecular substance finds a region of the RNA+ template on which it can be adsorbed in appropriate fashion, it may modify the pattern of the protein synthesized so that it possesses a specific enzymic relation to the substrate in question.

As long as the inducer remains attached and the template remains functional, the modified protein (adaptive enzyme) will go on being produced. The new template will not, however, be self-replicating. On the hypothesis that the protein will serve as a template for the RNA+ structure when this needs to be replicated (see page 29), we would present the cell with the problem of creating a replica (in whatever is the code of the RNA+ template) of something of quite different character. We postulate that in most instances the attempt to do so will fail and that success will need reorganization of some phase of the process by which the central genetic mechanism exerts an overall control. One knows, for instance, that effective temporary modification of the template RNA+ by inducer is only possible in bacterial strains of the right genetic constitution. Whether in some cases the occurrence of such modification of the RNA+ template results in the more frequent occurrence of the necessary gene mutation to allow the production of a constitutive enzyme is a matter for future discussion.

It is possible that the hypothesis of adsorption of inducer to the RNA+ template might provide implications susceptible of experimental test. The rather extraordinary behaviour of *Bacillus cereus* in regard to penicillin fits in very well with the modern view that penicillin interferes with the growth of susceptible strains by interference with the RNA-based mechanism of protein synthesis. In other words, where penicillin on being attached to RNA+ finds this genetically suitable it can provoke the appearance of penicillinase and so protect other RNA+ mechanisms. Where this is not possible there is a direct interference with the functioning of the protein-synthesizing mechanism.

A weakness of the hypothesis is that it provides no basis for the requirement that when a preparation such as that used by Gale and Folkes (1954) (sonically disintegrated staphylococci) is stimulated to produce β-galactosidase there is an associated synthesis of new RNA. Both processes are inhibited in parallel by penicillin. One can only suggest that whenever a RNA+ unit is functioning in a *new* fashion there is an automatic call for the production of more such units.

CHAPTER II

ANTIBODY PRODUCTION

When a man or a rabbit is inoculated with a bacterial vaccine or some other type of foreign organic material, a corresponding antibody appears in the circulating blood in 7–10 days time. This antibody is a modified gamma globulin, molecules of which have a specific complementary pattern (SCP) relationship to one or other of determinant groups on the foreign material (antigen). Antibody production is probably the best known and most extensively investigated of all the situations in which cells can be induced to produce proteins with functional patterns different from those that could be produced in the absence of the specific stimulus.

The basis of antibody production has been widely discussed since the days of the side-chain theory of Ehrlich. Two recent extensive discussions are those of Burnet and Fenner (1949) and Haurowitz (1952*b*), which may be taken as representative of two different approaches, the first with a strong biological bias, the second with an equally strong tendency toward a predominantly chemical approach. It would be inappropriate in the present context to give a summary of the experimental work prior to 1950 on which these two discussions are based. For the most part it will be convenient to develop this section on antibody production as if it were a continuation of the argument in Burnet and Fenner's monograph.

1. *The self-marker concept*

The essential features of their 'self-marker' theory of antibody production may be indicated by quoting from their summary:

'(1) The basis of our account is the recognition that the same system of cells is concerned both in the disposal of effete body cells (without antibody response) and of foreign organic material (with antibody response).

'(2) In order to allow this differentiation of function, expendable body cells carry self-marker components which allow recognition of their "self" character. Antigens in general are substances of the same chemical nature as the marker components but of different molecular configurations.'

Based largely on the existence of bovine twins in which two distinct blood groups can co-exist in the same individual and on Traub's work with congenital infection by lymphocytic choriomeningitis virus in mice, it was postulated that the self-recognition system was developed during embryonic life. Any antigenic pattern that reached what we called the scavenging cell system (reticulo-endothelial and associated cells) before a certain critical point around the time of birth or hatching, would be accepted as 'self' and in subsequent life its re-entry into the body would not provoke antibody production. This hypothesis, of course, suggested a number of possibilities for direct experimental test by the artificial introduction of antigens into embryonic animals. Our own experiments with influenza virus in chick embryos were, however, negative, probably because the virus could not be adequately disseminated in the embryo without killing it.

In the last five or six years, however, the correctness of the approach has been confirmed by several groups of workers. The most striking demonstration was that of Billingham, Brent and Medawar (1953) who found that by inoculating mice embryos *in utero* with a cellular emulsion of organs from another homozygous strain, they could be made to acquire a specific tolerance to skin grafts from this strain which persisted through their free living life. The 'foreign' skin is accepted as 'self' because of some modification of the host cells during the last stage of embryonic life. A point of some importance is that, whereas all the host's own cells and their descendants show this tolerance, other acceptable cells (from animals homozygous to the host) cannot acquire the tolerance, when subsequently introduced into the body. If a tolerant host of the white strain A carrying the black skin graft of the foreign type is grafted with a lymph node of a normal A mouse, the black graft begins to break down after a few weeks and is eliminated (Billingham *et al.* 1955). There could hardly be a more definite proof that in the limited sense of cell to cell inheritance within the body an inheritable change has been induced in the cells by foreign antigens.

Another set of experiments on similar lines is due to Buxton, (1954), who showed that a specific tolerance to a *Salmonella pullorum* vaccine could be induced by intravenous inoculation of the antigen in chick embryos between the twelfth and seventeenth day of incubation. Intravenous inoculation at the twentieth day, i.e. one day before hatching, or yolk sac inoculation at earlier stages both failed to influence subsequent response to the antigen. This underlines Medawar's statement that to produce specific immunological tolerance the antigen must be present at the right time, in adequate amount appropriately distributed in the body, and perhaps persisting for a sufficient period.

A successful attempt to reproduce experimentally the conditions in bovine twins is briefly reported by Horowitz and Owen (1954). They describe unpublished experiments by Ripley and Owen in which persistent erythrocytic mosaicism was produced by injecting rat embryos with embryonic rat cells from a donor of different blood type.

Further interesting work on the natural situations arising during pregnancy has been reported. In bovines Owen (1945) showed that it was frequent for non-identical twins to show two co-existing types of red cell in both circulations, one corresponding to the genotype of one twin, the other to that of the second. This mosaic condition or blood group chimæra persists throughout the lives of the animals. A similar condition has recently been described for sheep twins by Stormont *et al.* (1953).

An important extension of the same principle was made by Anderson *et al.* (1951), when they found that a similar mutual tolerance extends to the acceptance of skin grafts between non-identical bovine twins. Full brothers or sisters from separate pregnancies in cattle show the usual failure to accept such skin grafts. Essentially this finding shows that Nature can provide precisely the same situation that Billingham *et al.* produced experimentally in mice.

In cattle binovular twin pregnancies show a fused placental circulation, which allows relatively free interchange of soluble and cellular blood constituents. The results that have been described presumably depend on the transfer of hæmopoietic cells from one embryo to the other through the common circulation and their establishment in the bone marrow.

In human beings the placental circulations of binovular twins are characteristically separate. In fact, this is one of the classical criteria for differentiating binovular from monovular twins. Very rarely, however, the rule fails to hold and a healthy woman was recently described by Dunsford *et al.* (1953) whose blood contained nearly equal numbers of Group *A* and Group *O* cells. The two types of red cell could be separated and their other serological reactions tested. The qualities in which the cells differed were as follows *O*, *kk*, $Jk(a+b+)$; *A*, *Kk*, $Jk(a-b+)$.

The subject was a secretor of *O*(*H*) substance and there can be no doubt that the *O* cells were the subject's 'own'; the others were those of her twin brother who had died as an infant 25 years previously.

There has been considerable interest in the question of why only some of the women of *Rh*-negative type who produce

Rh-positive offspring are immunized by the *D* antigen, and why only some of these show severe symptoms in the infant immediately after birth. In an examination of serological data, irrespective of whether or not the child gave any clinical evidence of erythroblastosis, Owen *et al.* (1954) found that the *Rh* status of the woman's mother was significant. Women *Rh*-negative who were married to *Rh*-positive husbands and had experienced three pregnancies were divided into those who had *Rh*-positive and *Rk*-negative mothers. Amongst 29 who had *Rh*-positive mothers, 25 had not produced antibody (86 %) while of 22 with *Rh*-negative mothers only 9 had failed to produce antibody (41 %). The difference is significant and suggests that many of the first group had been conditioned to tolerate *Rh*-positive antigen by exposure to it *in utero*. No influence, however, could be recognized on the clinical manifestation in the infants. A similar negative result had previously been reported from England by Booth *et al.* (1953).

The existence and specificity of acquired pre-natal tolerance is now well established and it is clear that its interpretation must find a place in any theoretical discussion of antibody production.

2. *Antibody production after the elimination of antigen*

The second feature of Burnet and Fenner's formulation on which there was and is considerable difference of opinion at the experimental level was the claim that 'the production of intracellular units and their partial replicas (antibody) may continue for long periods after the antigen has disappeared from the body. Their activity and specificity steadily decline in the absence of a fresh stimulus by the same antigen but . . . with further such stimuli there is accelerated activity and the possibility of further change in the specificity of the antibody produced.'

This claim can probably be regarded as defining the essential difference between Burnet and Fenner's approach and the orthodox formulation to which the names of Haurowitz, Mudd and Pauling are particularly attached. Their basic hypothesis is that each antibody molecule has its specificity determined by physical contact with antigenic determinants retained in the antibody-producing cells. Antibody production will, therefore, be impossible when the antigen (or its significant groupings) is eliminated completely.

This is a question which at first sight appears to be ideally suited for solution by the use of appropriately labelled antigens. Many experiments along these lines have been reported, but the results have not been easy to interpret. Haurowitz (1952*b*) finds that the results are still compatible with the view that antigen persists in the body throughout the period during which antibody production can occur. Others are equally convinced that the evidence decisively favours the view that antibody production can continue in the absence of antigen.

Recent experiments bearing on the problem can be divided into those in which the antigenic label takes the form of (i) a dye or other recognizable group attached to the protein antigen by diazotization, or (ii) an isotopic label of which I^{132} seems to be the most suited to the purpose and is certainly the one most extensively used. A third approach due to Coons and Kaplan (1950) is to trace the antigen by using fluorescent antibody as a histochemical reagent.

If one injects a foreign protein, e.g. bovine gamma globulin, intravenously in a normal rabbit, the fall in the concentration as measured, e.g. by an isotopic label or by a quantitative immunological method, is in three phases. In the first 24 hours there is a fall to approximately 50 % due to redistribution of the protein between blood and tissues. Then a fall at a slower rate with an indicated half life of about 4·6 days till the seventh to ninth day when, with the appearance of antibody, there is a precipitate disappearance of antigen from the circulation. The antigen is taken up in the tissues largely in macrophage cells of the reticulo-endothelial system, but by no means exclusively so. Lightly labelled material gives no evidence of being stored in any tissue (Dixon *et al.* 1951, 1953), but azo-proteins behave quite differently and are stored for long periods. Kruse and McMaster (1949) detected an ⌜azo dye, globulin⌝ antigen in mice up to 120 days after injection and claimed that native antigen could also be detected by an unorthodox method over the same period. Ingraham (1951), however, using an azo-sulphanilo globulin labelled with S^{35}, found that the label was stored for 200 days in liver and spleen but no specific antigenic activity could be detected in the liver after 17 days.

Similar long persistence of antigenic material is well known to occur with pneumococcal polysaccharides. The nature of the immunological paralysis produced in mice by large doses has been discussed in the monograph by Burnet and Fenner (1949), and is referred to again on p. 77 below.

At the risk of oversimplification we believe that the examples in which long persistence can be demonstrated have all involved highly abnormal substances as compared to the pathogenic micro-organisms and the body's own cells which are the agents that have been biologically relevant to the evolution of the immune mechanism. Even purified pneumococcal polysaccharide is something considerably different from the complex with lipid and protein in which it occurs in the bacterial surface. As will become more evident later, it seems almost irrelevant to the problem of the antigen as template whether or not intact antigen can be demonstrated in tissue. It is quite inconceivable that a whole foreign macromolecule could be built into one protein synthetic unit and go on impressing a complementary specificity on each antibody molecule being produced for years and years. It would fit the facts far better to see these persistent antigens as abnormal agents which can be broken down by body enzymes only with the greatest difficulty so that there is only a minimal liberation of soluble material which is, however, enough

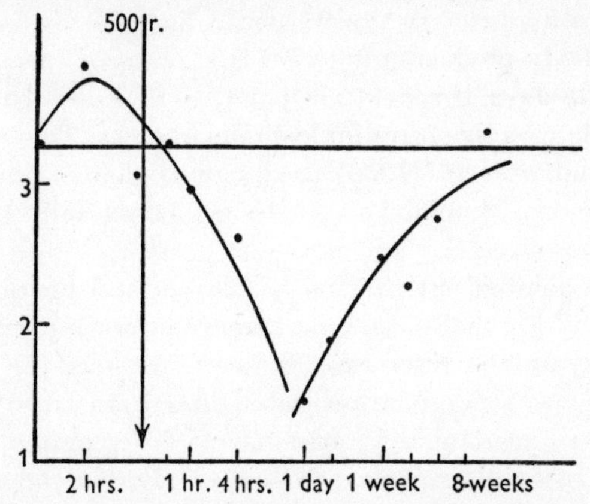

Fig. 3. The influence of X-irradiation on antibody response. The antibody response is shown (ordinates) to a standard dose of antigen given intravenously before and after a single whole-body irradiation of 500 r. at time zero. The time scale is approximately logarithmic from 1 hr. to 8 weeks. (Data from Taliaferro and Taliaferro, 1954.)

to maintain the antigenic stimulus in the case of pneumococcal polysaccharide.

Probably the most impressive evidence against the importance of persisting antigen is provided by the results of X-ray treatment. It is known that X-irradiation in adequate dose shortly before administration of an antigen will prevent antibody production or greatly reduce the yield. This phenomenon has been studied in detail by Taliaferro and his collaborators (1952, 1954). They were concerned with the effect of X-rays on the ability of rabbits to produce hæmolysin in response to a single intravenous injection of sheep red cells. From the present point of view the most important finding was that when the antigen injection was given 24 hr. after irradiation no trace of antibody appeared in 9 out of 10 rabbits. When these were subsequently tested with the same antigen more than a month later, they responded like a normal animal with a primary type response. Since X-ray damage to the antibody-producing function is almost completely overcome in 28 days, this evidence points to the persistence of the antigen in effective form for less than 4 weeks. Talmage *et al.* (1951) and Stevens (1953) used bovine gamma globulin in experiments of similar type, both with essentially the same results.

As we pointed out previously (Burnet and Fenner, 1949) the crux of the matter is the difference between primary and secondary or later responses. In many instances the primary response may seem to be associated directly with the presence of antigen, but later responses have a different quality. The antigen is rapidly removed from the body, but the antibody response is both more rapid and much more prolonged. Dixon *et al.* (1954) showed that production of antibody could continue for more than six months after a tertiary antigenic stimulus of bovine serum globulin or albumin.

In the 1949 monograph we laid much stress on the evidence that immunity from virus diseases like measles and yellow fever could persist for up to 50 years without any fresh contact with the virus. This has not been extensively discussed by others in recent years, but it is likely that most workers feel that the possibility of persisting latent infection must always weaken such arguments. Nothing has emerged in recent work, however, to modify our opinion that the classical epidemiological study of measles in the Faroë Islands by Panum (1847) still provides very strong evidence for long-lasting immunity in the absence of persisting infection.

3. *The site of antibody production*

When we turn to the question of the site of antibody formation, most recent work seems to have confirmed the general point of view adopted in 1949 (Burnet and Fenner, p. 129): 'The antibody producing mechanism is initiated by the entry of the antigen into phagocytic cells of the reticulo-endothelial system (macrophages) . . . antibody producing units are transferred to reticulum cells or other relatively undifferentiated mesenchymal cells in the immediate vicinity of the macrophages. Under conditions inducing active antibody formation, these cells multiply freely and take on the staining qualities of plasma cells. They are responsible for the actual production of antibody during the peak phase of the response. The reticulum cells also give rise to lymphocytes which are probably responsible for maintenance of low levels of antibody long after the antigenic stimulus.'

Of the new evidence available since 1949–50, the most important contributions are concerned with the influence of X-rays on the immune response and the possibilities of transferring antibody-producing power by cells or tissue grafts.

The work of Taliaferro *et al.* (1952) on the effect of X-rays on hæmolysin production in rabbits has already been briefly referred to. They showed that if the antigen was given before irradiation there was little or no effect except for a very short period about 2 hr. after the injection of antigen when a definite *increase* in antibody production was noted. When antigen is given after X-irradiation with a standard dose (500 r. whole body irradiation) the effect increases very rapidly, reaching a peak (i.e. minimal antibody production) at 24 hr. The normal induction period of the antibody response and rise to peak are both lengthened as compared with the control values which were 3·6 and 5·2 days respectively. In rabbits inoculated 2 days after irradiation the values were 19 and 10·5 days. Inoculation 2 days before irradiation gave values of 7·2 and 9·5 days and, at 2 hr. after irradiation when the stimulation effect was shown, values of 7·5 and 10·5 days, both very significantly longer than the controls. In 9 out of 10 instances there was no antibody response at 24 hr. after irradiation.

It seems then that the X-ray lesion takes some hours to develop and is at its full intensity at 24–48 hr. thereafter, gradually returning to normal, which is reached in 4–8 weeks. With a single intravenous injection as antigenic stimulus the key process which is interfered with must occur within a few hours of injection and must be completed within a further few hours.

Maurer *et al.* (1953) followed the associated histological changes and noted particularly the almost complete disappearance of small lymphocytes from the body. The lymph nodes at the height of the reaction were depleted of lymphocytes leaving a loose network of reticulo-endothelial cells and fibres in which were scattered large lymphocytes, plasma cells and macrophages and fibroblasts. The second phase of antibody synthesis presumably takes place in some of these radio-resistant cells.

Harris *et al.* (1954*a*) have used the passive transfer of cells from rabbit lymph nodes to transfer immunity in rabbits. In order to avoid the possibility that antibody was being produced actively by the recipient in response to antigen present in the lymph node extracts, they used recipient rabbits which had been X-irradiated 24 hr. before receiving the cell suspensions. Harris *et al.* were chiefly interested in the first stages of the process and transferred popliteal lymph node extracts at short intervals, 10 minutes to 4 days, after inoculation of killed dysentery bacilli into the corresponding foot. They found that, virtually irrespective of whether the cells were in the donor or recipient animals, antibody production sufficient to be detected in the recipient's serum appeared 4 days after the inoculation of antigen into the donor. The results are rendered difficult to interpret, however, by the subsequent finding that, even if antigen is incubated *in vitro* with donor lymph node cells or injected simultaneously, an essentially similar response results (Harris *et al.* 1954*b*). If, as the work of Cole *et al.* (1953) suggests, this is due to the non-irradiated cells being able to provide a soluble or particulate replacement of something destroyed by X-ray in the recipient, some of the results are conceivably due to active immunization. The results of experiments in which the lymph nodes are removed 4 days after inoculation and give rise to antibody in the recipient within 24 hr., however, can hardly be so interpreted. Their chief significance is to establish with reasonable certainty that the whole process of antibody production can take place in the lymph node, thus confirming many earlier investigations along different lines.

The most direct approach to the type of cell actually producing antibody is to apply a histochemical test for antibody in or on the cell. Coons *et al.* (1953) used a modification of the fluorescent antibody technique which they call the 'sandwich' method. By treating sections of lymph nodes from animals immunized with an antigen A, first with that antigen and then with the corresponding fluorescent antibody, they built up the system ⌜fixed cellular antibody, antigen, fluorescent antibody⌝. After first showing that no antigen could be detected by the fluorescent antibody technique in lymph node cells after the very early stages of immunization, they were able to accept areas showing fluorescence after combined treatment as being associated with cells containing incorporated antibody and, therefore, being presumably responsible for antibody production. In their experiments the first cells to show up were scattered in the medullary cords. They had large nuclei and a thin rim of cytoplasm which apparently contained a high antibody concentration. These could well be early derivatives of reticulum cells. Two to four days later the stained cells were of early plasma cell type. When a secondary response was studied in the same way, antibody was first detectable on the second day in cells of younger type. Many more cells were involved and they were grouped in colonies suggesting that the descendants of some scattered primitive cells were now all engaged in antibody synthesis.

In another study of the cellular morphology of antibody production, White *et al.* (1953) studied the local lesion produced by alum-precipitated antigen. The cell types containing antibody were similar to those observed in the lymph nodes by Coons *et al.* (1953). Primitive cells with the morphology of hæmocytoblasts, immature and mature plasma cells showed fluorescence. The macrophages contained granules of aluminium hydroxide, but were devoid of antibody.

Reiss *et al.* (1950) examined the capacity of cells from the popliteal lymph nodes of immunized rabbits to adsorb the specific bacterial antigen. Immature plasma cells were pre-eminently active, but mature plasma cells only irregularly, while small lymphocytes were inert.

Taking all the evidence into account there can be little hesitation in adopting the view of Bjornboe *et al.* (1947) and Fagreus (1948) that cells of the plasma cell series are the predominant producers of antibody in the acute phase of the immune response.

The function of lymphocyte is not quite so clear. Hayes and Dougherty (1954), using a technique involving lymphocytes in the mouse peritoneal cavity, found that in immunized animals the homologous bacterial antigen was agglutinated against the surface of lymphocytes and not of other cells.

The suggestion that immune lymphocytes must be actively functioning if they were to show evidence of antibody production is supported by the work of Wesslen (1952). He collected lymphocytes from the thoracic duct of rabbits that had been immunized against *Salmonella typhi* some weeks previously. The cells in the thoracic duct lymph are almost exclusively small lymphocytes. If these were washed with tyrode solution and then ground with distilled water, the extract contained no agglutinin. If, however, they were maintained under tissue culture conditions in tyrode solution for 8–12 hr. at 37°C., considerable amounts of specific immune agglutinin were liberated into the fluid. If this result is confirmed, it would seem to indicate that antibody is liberated by lymphocytes as soon as it is synthesized.

At various times there have been suggestions that other types of cell than those of the reticulo-endothelial and lymphoid series were concerned in antibody production. Haurowitz (1952*b*), for instance, believes that antibodies can be produced in all cells in which proteins are formed, provided the cells are capable of binding molecules of injected antigen. In their study of the distribution of iodinated antigens in liver fractions, Haurowitz *et al.* (1952) found that except for the first few minutes when the microsome fraction contained the largest proportion, most of the radioactivity was associated with the mitochrondria. The liver, however, is not a site of γ-globulin synthesis (Miller *et al.* 1954) nor, from unpublished investigations cited by Coons (1954), is antibody produced there.

The conclusion seems justified that antibody production is a specialized function of mesenchymal cells and not something that is common to any type of vertebrate cell.

4. *Theoretical approach to antibody production*

There has been no experimental development to invalidate the point of view adopted in 1949, but the recent clarification of the process of adaptive enzyme formation will obviously call for a restatement of the analogy that was drawn between antibody production and adaptive enzyme formation. In view of the strong current opinion that, at least in the great majority of instances, adaptive enzymes are not produced by self-replicating mechanism, special attention must be given to the question of the transmission of an antibody-producing capacity from a cell to its descendants. In our previous formulation we laid stress on the fact that 'antibody production is a function not only of the cells originally stimulated but of their descendants'.

This is a somewhat heretical point of view which, however, does not seem to have been discussed favourably or unfavourably by others interested in the problem. If lymphocytes are the major transporters and liberators of antibody, and if, as is the current teaching, the average life of a lymphocyte is much less than 24 hr., we are compelled to accept the point of view or produce some elaborate *ad hoc* explanation to account for the transfer of antibody-producing capacity from the 'real' antibody-producers. Strong and rather direct support for this opinion is given by Coons's (1954) finding that when a secondary response is initiated, antibody-containing cells are distributed in the medullary cords of the lymph nodes in patterns that point to their having arisen from the multiplication of a relatively small number of parent cells scattered through the tissue.

If we accept this limited inheritance of antibody-producing capacity, there are two ways in which it can be mediated.

(1) The whole genetic apparatus of the cell may be modified so that the new pattern of protein produced is a result of a chromosomal genetic change equivalent to a directed mutation.

(2) The inheritance is cytoplasmic in character and is transmitted from one cell to another in ways which resemble such processes as the distribution of plastids in plant cells or of melanin granules in the chromatophores of the guinea-pig's skin.

We have always maintained that the second alternative is much the more likely, but we are now inclined to think that some important aspects of 'cytoplasmic' inheritance may be related to the nucleus though not to the conventional genetic mechanism of the chromosomes. This is based in part on Caspersson's contention that RNA is largely synthesized in the nucleus and passes thence to the cytoplasm, and in part on Coons's very consistent finding of antigens and viruses in the nucleus at an early stage of their entry into cells (Coons *et al.* 1951).

We may now attempt to combine earlier data with the work on theoretical immunology since 1949, which has been discussed in the preceding sections, to produce a summary of the essential characteristics of the antibody response that must be covered by any acceptable formulation of the process.

(1) Self-components which are antigens for other individuals or species are not antigenic for the animal itself.

(2) But proteins from some non-expendable tissues, such as the lens of the eye, may be exceptions to (1).

(3) All antigens are proteins or complexes of an antigenic determinant with protein, which in some instances may be provided by the animal immunized or sensitized.

(4) Specific antigenic pattern may be expressed in protein, polysaccharide and probably in nucleic acid (Blix *et al.* 1954), as well as in small molecular form.

(5) There are patterns which are much 'better' antigens than others of the same general quality; some proteins are non-antigenic.

(6) Various types of haptenes exist which can react with antibody but not provoke its production. Two significant examples are Pz protein of *E. coli* and virus protein from turnip yellows.

(7) Embryonic animals do not produce antibody.

(8) Appropriate administration of antigen at some periods of embryonic life results in an acquired specific tolerance of the antigen so that in later life the animal will not produce antibody to it.

(9) Irradiation with an adequate dose of X-rays within a few days before primary injection of antigen prevents antibody response. Irradiation 2 or more days after injection does not interfere with the development of the response.

(10) For most and perhaps all antigens there is a striking difference between the primary and the secondary response; the latter comes on much more rapidly, reaches a higher titre and lasts much longer than the primary response.

(11) There are inheritable differences, between individual animals, in readiness of antibody response and in the type of antibody produced in response to the same antigen.

(12) The type of antibody produced also varies according to the species and age of the animal used and according to the time relations of the antigenic stimuli applied.

(13) In addition to the production of classical antibody at least two other types of immunological response exist—'hay fever' type and 'tuberculin' type sensitization.

(14) Antigens are taken up predominantly by macrophages of the reticulo-endothelial system—not by plasma cells or lymphocytes.

(15) In the early stages, antigen may often be detected by Coons's method in the nucleus of reticulo-endothelial cells.

(16) Antibody is probably produced exclusively by cells of the ⌜plasma cell, lymphocytic⌝ series—not by reticulo-endothelial cells.

(17) Cells actively producing antibody show the presence of increased RNA in the cytoplasm; it is this which defines a plasma cell.

(18) Antibody is not a hydrolytic or depolymerizing enzyme.

(19) Antibody is or can be produced after all antigen in any recognizable form has been eliminated from the body or destroyed.

(20) Antibody must be produced by descendants of the cells primarily involved.

Lack of antigenicity of self-components. We still regard the absence of antigenicity of the body's own components as the most important feature to be interpreted and will commence the discussion with a restatement of the 'self-marker' hypothesis. In one way or another there must be a means of recognizing the difference between 'self' and 'not-self' material. There appears to be no limit to the number of molecular patterns that can act as antigenic determinants. On the other hand, there are hints that only a relatively small number of configurations are concerned in 'self-recognition'. It is clearly economical of hypothesis to assume that there is a positive function of self-recognition that is followed by what can be called an 'instruction' that the antibody-producing function be inhibited. Without such a positive instruction the intrusion of organic material into scavenger cells is followed by antibody production to some degree.

On this view, the first problem to be considered is what determines that foreign and worn-out material is taken up by macrophages while normal cells and apparently some micro-organisms such as the viruses of serum hepatitis and equine infectious anæmia are not. The function of polymorphonuclear leucocytes would also have to be considered here. But simply because it is currently unorthodox to pay much attention to this first step in immunological discussion, and because there is little modern work on the subject, nothing more will be said.

With the entry of foreign organic material or damaged and outdated body cells into the scavenger cells of the reticulo-endothelial system we are faced with the necessity of devising, on the basis of other biological analogies and of the facts of immunology, the most probable picture of the recognition mechanism that we have to postulate. It is the thesis of this monograph that, where biological matters above a certain level of complexity are concerned, most interpretations must be in terms of macromolecular pattern which, by interacting with complementary or near complementary pattern in some other functional situation, can induce action or, if it is more convenient so to express it, convey information or instructions. The only mechanism we can conceive, therefore, for recognition is a series of complementary structures to a rather small number of 'self-markers'. If we forestall parts of the subsequent discussion, we may perhaps define self-markers as those chemical configurations which, given the genetically determined chemical structure of the body, would otherwise be capable of acting as antigens. This is in many ways a confession of ignorance, but we are aware that there are many proteins—gelatin, for instance—that are altogether non-antigenic. We know, too, that after a certain experience of passage there are transplantable neoplasms which are tolerated in a way that no normal tissue from the same (heterozygous) donor would be.

We assume that there are in any given scavenger cell a population of protein molecules of perhaps 4–10 types carrying complementary configurations to the self-marker groups. The possibility of a 'master key' carried on one molecule which would find ways of fitting all the 'locks' presented by several self-markers would also have to be considered. One would guess that these proteins were globulins with a general resemblance to the γ-globulin of the serum. Such proteins are known to be present in cells of various types (Gitlin, Landing and Whipple, 1953). The relationship of the protein which we shall call a *Recognition Unit* (RU) to the self-marker will be analogous to that of antibody to antigen or perhaps of enzyme to substrate. It is impossible to know what the precise function of that union is. We should guess that in essence it is to keep the potentially antigenic determinant (the self-marker) under control until, by the enzymic breakdown of its cellular or protein carrier, it has lost any antigenic potentialities. This, of course, merely underlines our ignorance of why a protein carrier is necessary before a determinant group or haptene will function antigenically. The other possibility is that some or all of the RU's are enzymes which actively destroy the self-markers and, in this way, render the complexes non-antigenic.

The existence of acquired specific tolerance to potential antigens encountered before a critical point in embryonic development makes it more than probable that the self-marker mechanism is laid down by a non-genetic mechanism during embryonic life. The production of RU's in embryonic life is the protype of antibody production in the fully developed animal and any theoretical picture must be developed with that in mind.

There are extensive hints discussed by Medawar (1947) that at various stages of development particulates pass from one cell to another and, in one way or another, influence the development of the recipient cell. Quite early in development, expendable cells must arise and, simultaneously, a means of dealing with them when they are worn out. We should guess that something takes place equivalent to the *Entwicklungsmechanik* of enzyme formation by substrate that has been outlined by Pollock. Using, perhaps unjustifiably, all the hints that can be drawn from the process of adaptive enzyme development, one can picture that the effective agent (the potential self-marker) is the partial breakdown product of the effete cell which can intrude successfully into one of the protein-synthesizing mechanisms of the cell.

Reverting to our picture of protein synthesis in which a

template of RNA + supplies the pattern and the energy for synthesis, we may assume that, just as penicillin (or the significant fragment of that molecule) by intruding into the RNA + template gives rise to the synthesis of the adaptive enzyme penicillinase, so the significant fragment of the self-marker may be similarly incorporated. The result could be the production of a protein with a new pattern complementary to the self-marker, a pattern which according to circumstances might confer enzymic activity or not. It would be wholly in line with evolutionary principles if a mechanism primitively developed for enzyme production should be deflected in this manner to a new use. The presence of such modified protein in the cell would tend to protect other protein-synthesizing units from interference in this fashion. In order to allow them to proceed with their proper activities it would be highly desirable to concentrate the protective function on one particular set of synthetic units.

To provide a speculative picture of how recognition units might be developed to fulfil such a function, we can conceive that intracellular γ-globulin producers are primarily concerned. Using the same concept that RNA + and protein can act as mutual templates under suitable conditions for synthesis of the other, we can imagine the steps shown in Fig. 4. The self-marker is caught into the functional aspect of a template RNA +, and the protein synthesized from this will contain a new specific complementary pattern (SCP) related to the self-marker. If now the new protein in its turn serves as a template for RNA + synthesis we have the possibility that what we shall term a 'genocopy' of the ⌜RNA +, self-marker⌝ complex will be produced. This is a template built up in normal fashion, which can confer the same specificity to the protein produced as the complex would.

A mechanical model for this concept might be found in a mould made of fusible alloy which is used to make a plaster of paris cast: the mould initially produces a smooth spherical cast; it is desired, however, to have an impressed decoration on the cast; this can be arranged by attaching to the inner surface of the mould a complementary raised pattern in dental wax or plasticine; this will give the decorated cast that is wanted. Now by the use of appropriate techniques, it will be possible to use the decorated cast as a pattern around which fusible alloy can be poured to form a new mould in

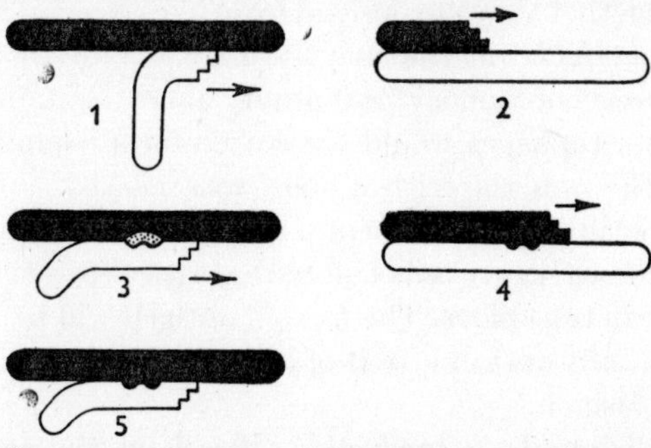

FIG. 4. A diagram to illustrate the 'genocopy' idea in the synthesis of antibody. 1, 2. Normal synthesis of protein-RNA+ templates. 3. Intrusion of antigenic determinant (stippled) with synthesis of modified protein. 4. Synthesis of 'genocopy' of determinant against modified protein template. 5. Definitive synthesis of antibody protein.

which the complementary pattern is expressed solely in metal. This is the genocopy of the mould modified with wax or plasticine.

It would seem that γ-globulins are in some way particularly apt for this type of interaction between templates. Details of the process must be left vague, but it would be in line with Caspersson's ideas if the synthesis of the genocopies (RNA+) took place in the nucleus and that these then passed to the cytoplasm where most or all of the protein production takes place. This will result in the presence of specifically patterned RU's in the cell which by hypothesis will react with organic material bearing the corresponding self-marker and so facilitate its effective disposal. If, however, a different type of marker reaches the cell, it will be able to intrude into the synthetic processes and set going the same sequence giving rise eventually to another type of RU.

All this still, of course, refers to embryonic life. The question of the origin of macrophages is undecided, but there appears to be no substantial reason for not making the simplest assumption that they are derived from like cells and may take on either a circulating (monocyte) or tissue form (histiocyte). By movement of monocytes it would almost certainly result that all macrophages would have a uniform complement of RU's by the time the critical point was reached.

If, by natural or experimental processes, a foreign antigen reaches the scavenger cells before the critical point, a similar reaction will take place. The foreign antigen will be regarded as a potentially antigenic part of the self and dealt with in the standard fashion.

Nature of antibody production. Based on the concept of recognition units (RU) developed in the previous section, the simplest interpretation of antibody production is to imagine precisely the same reactions, but set at a higher intensity and with some qualifying factors arising from the nature of the evolutionary situation with which it is concerned.

Briefly, a foreign antigen enters a scavenger cell and foreign determinant groups, unblocked by RU's, are liberated in a situation where they can be incorporated into developing RNA+ templates, perhaps in the nucleus. The evidence is compatible with the view that the immediate response to a primary antigenic stimulus, in which only one effective macromolecule reaches the cell being considered, is limited to the construction of the primary template and the synthesis of a minimal number of protein molecules analogous to recognition units but which we can now call AU's (*Antibody Units* in intracellular form).

Subsequent stages must form a sequence of processes which at some point involves a transference of activity from cells of the reticulo-endothelial series to those of the lymphoid series. At what stage this transfer takes place is not known. It will be less confusing, however, in developing a theoretical approach, to make some specific assumptions on the matter with the understanding that such development is justified only as a frame on which to form working hypotheses. We should guess then (i) that the production of the primary template and its first genocopy takes place in the reticulo-endothelial cell, (ii) that the synthetic units, genocopy RNA + with or without other components, are transferred to those stem cells of the lymphoid series that are in the immediate neighbourhood of the macrophage primarily involved, and (iii) that the units are then incorporated as part of the protein synthetic mechanism of the stem cell—perhaps again in the nucleus.

The production of classical circulating antibody is a result of the active synthetic process in cells which because of their synthetic activity have the staining qualities of immature plasma cells, and liberation of the antibody molecules into lymph spaces or the blood sinuses of the spleen or bone marrow. There is much to suggest that all liberation of antibody is part of a secondary process. With the laying-down of the antibody-producing mechanism we have also to picture the installation of some sort of trigger mechanism by which the re-entry of the specific antigen into the body will set going a process of active protein synthesis and cellular replication in the stem cells carrying the transferred AU-producing mechanism. Once liberated, antibody molecules become simply part of the expendable protein of the blood plasma with a half life of 4–20 days depending on the species of animal involved. Its function as a protective agent depends simply on the power of an antibody molecule to combine firmly, by its SCP relationship, with soluble or particulate antigen which may enter the body. This, however, is an aspect of immunology with which we are here not concerned.

On this view antibody production with liberation into the body fluids is always the result of a secondary antigenic stimulus. This need not, however, be due to a second injection of the antigen or a second contact with the infecting micro-organism. After almost any type of immunizing injection the antigen will form a deposit somewhere in the tissues from which antigenic molecules will be progressively released. Especially when the antigen is in the form of relatively large particles such as bacteria or viruses, or is segregated into almost insoluble granules as is often the case with azo-dye proteins, release of active antigenic material may go on over a period of weeks, months or even years. Almost always, therefore, there will be a stage when antigen released from the primary deposit will reach cells in which the antibody-producing capacity has already been laid down and provoke them to active synthesis and liberation of antibody.

This briefly is one way of envisaging the production of antibody in its classical circulating form. There are other types of immunological response that are concerned with a variety of sensitizing reactions and with the response to grafting of foreign tissues. In view of the importance of the situations in which 'self-markers' appear to be changed by appropriate administration of chemical substances, it is more convenient to discuss in separate sections the sensitization reactions associated with simple chemical substances, and Green's ideas of the part played by immunological factors in chemical carcinogenesis. Brief reference will also be made later to the 'hay-fever' type of antibody that is produced under certain conditions.

This formulation of antibody production differs from that of Burnet and Fenner's (1949) presentation only in the adoption of a theory of protein biosynthesis more in line with recent work on adaptive enzymes. In the earlier account it was assumed simply that the AU's were adaptive enzymes which had been modified from RU's of the same general character. A RU, for instance, which corresponded to a polysaccharide self-marker would be modified to an AU by any foreign polysaccharide of the same general type. Once the AU had been produced it was regarded as a self-replicating unit capable of transfer to a nurse-cell in which it could also multiply. Once they have been produced 'these secondary modified units have the normal capacity to replicate, accentuated and modified by the evolutionary development of their function as protectors against disease. According to circumstances they may be transferred to other cells or give rise to partial replicas carrying the specific complementary pattern but not the enzymic activity. These partial replicas are the globulin molecules of circulating antibody' (Burnet and Fenner, 1949, p. 106). The chief weakness of this formulation was the apparent insistence on the enzymic character of the process, although this was strongly qualified as representing merely the closest analogy we could find.

In the earlier account no attempt was made to particularize about the nature of protein biosynthesis, and when a replicating protein was spoken of, the mechanism responsible for the replication was left in the background. In the present account it is accepted that protein is produced by a RNA-containing template so that the recognition units and antibody-producing units of the earlier account represent the complementary system (protein and RNA+) of this, while 'partial replica' becomes protein or antibody molecule.

The most popular current theory of antibody production is that for which Haurowitz, Mudd and Alexander were initially responsible in one form or another. It was adopted by Pauling as a basis for his immunological work and its most elaborate recent exposition is due to Haurowitz (1952*b*).

Haurowitz's formulation is based on the hypothesis that protein synthesis is by replication of an elongated polypeptide chain model. The appropriate amino acids from the pool find their place by a process akin to crystallization, RNA serving merely as a basis to hold the model in stretched condition and in all probability to assist the process of linking the amino acid residues together. Once the model has been copied, the polypeptide chain is released, but as yet it has none of the specific functional character of the protein it will eventually become. In Haurowitz's view specificity is determined by a controlled folding of the polypeptide chain against a secondary template. The complementary pattern of antibody is produced by the folding of a non-specific γ-globulin type of chain against a template into which the antigen or its determinant group is incorporated.

This point of view is in many ways not very different from our own, particularly since, as seems likely, Haurowitz would now accept the suggestion that the remoulding of the polypeptide chain on the secondary template is, in part, a result of an appropriate set of transpeptidation replacements of one amino acid by another. Although most biochemists do not seem to be attracted to Haurowitz's model polypeptide, there is no formal reason why it should not represent the + in our RNA +. If we were obliged to develop a theory of antibody production on the basis of Haurowitz's formulation of protein biosynthesis, we should (i) endow the secondary template with a capacity to modify amino acid arrangement as well as the folding of the chain, (ii) provide a means by which the secondary template can be resynthesized using the final protein (perhaps in a penultimate state) as a template, so allowing the production of what we have called a 'genocopy' of the essential configuration of the antigen. Once these additions had been made, a theoretical formulation to cover all the facts in which we are interested could readily be developed.

The Haurowitz theory as it stands fails to account for the following aspects of the immune response: (i) the failure of embryonic animals to produce antibody; (ii) prenatally acquired specific tolerance; (iii) the characteristic difference of primary and secondary response; (iv) the persistence of antibody after antigen has disappeared from the body.

The fourth of these may still be the subject of legitimate doubt, but the first three are completely established as experimental and observational facts. No formulation of the Haurowitz-Pauling theory has yet attempted to account for them, and until this can be done its usefulness will be limited to the situations which it was originally invoked to explain, the reactions *in vitro* of antigen and antibody. It represents essentially a chemist's point of view and perhaps because of this fails to cover several aspects which the biologist finds of the greatest interest. It is quite certain that any alternative formulation of immunological theory will have to undergo periodic alterations, sometimes of a drastic nature, as information accumulates and new concepts in fundamental biology develop. The present restatement of the 'self-marker' approach differs considerably from that of 1949 and it may be desirable to conclude this section with an examination of how the present version accords with the list of features of antibody production and function given on p. 54 of the preceding section. Many of these have been covered already in the discussion.

(1), (2) The significance of 'self' and 'not-self' in relation to immunity is the basis of the whole approach, and hence the only point needing comment is the potential antigenicity of certain components of unexpendable tissues like the lens protein and perhaps some components of the central nervous system. Since the recognition of self is something 'learnt' during embryonic life and not genetically ingrained, it is to be expected that unexpendable tissues might have potential antigenicity. It should be experimentally feasible to show that if lens cells were injected intraperitoneally in fœtal rabbits they would develop incapacity to respond to lens antigens in later life.

(3) The necessity that, to act as an antigenic determinant, material must either be a protein or combined with a protein may possibly fail to hold for some of the bacterial polysaccharides. No satisfactory explanation has been given, but it is probable that macromolecular status is necessary to ensure that the self-marker or the antigenic determinant reaches the site of primary template formation which seems likely to be in the nucleus of the reticulo-endothelial cell.

(4), (5), (6) We contend that it is of the essence of the problem of macromolecular pattern relationships that orthodox chemical principles are inadequate to deal with questions of pattern transfer and replication. The hypothesis that the determinant group is actually incorporated into the RNA+ template of a globulin-synthesizing unit is legitimate even if subsequent evidence invalidates the idea that RNA has any template function. As long as any type of template hypothesis is accepted, only insignificant variations in the expression of the present theory would be needed. Whatever the nature of the template, it is obvious that some determinants will be capable of a better fit than others. Some may inactivate the whole function of the unit involved. It is probably by such essentially accidental relationships between the pattern of the unmodified template and the nature of the antigenic determinant that inheritable differences in immunizability (see (11), p. 72) depend.

The existence of proteins which are neither antigens nor haptenes depends presumably on the ease with which they are broken down into non-determinant fragments. The conversion of gelatin into an antigen by coupling glucosidyltrysine groups (Clinton, Harington and Yuill, 1938) confirms other hints that groups containing aromatic amino acids are common antigenic determinants. This may eventually become relevant to Schwartz's recent (1955) suggestion that the guiding function of a nucleic acid template in protein synthesis is primarily to determine the spacing of aromatic amino acids in the polypeptide chain.

The immunological significance of the Pz-Gz relationship, already discussed extensively in relation to the formation of adaptive β-galactosidase formation in *Escherichia coli*, is unknown, but its elucidation might be important for the understanding of protein biosynthesis in general. The simplest explanation would be to assume that a segment of protein A–A cannot function as a determinant, but if in the course of adaptive modification this becomes A–B, the segment can act as an antigenic determinant and produce anti-A patterns which can react with A–A as well as A–B.

Another point worth considering here is the influence of adjuvants in making some antigens, otherwise virtually inert, capable of inducing an effective antibody response. There may be several factors involved. The most important is probably the maintenance of a depot of antigen from which antigenic molecules are seeping out to the antibody-producing mechanism over a long period. Other factors may be concerned with the local accumulation of macrophages and other mesenchymal cells around the deposit of antigen. Holt (1949) suggested that macrophages might take up toxoid adsorbed on aluminium phosphate particles and transport them to various other sites for antibody production. Oakley, Batty and Warrack (1951) have shown that the cells surrounding a depot of alum-precipitated toxoid are themselves potent producers of antitoxin. Neither of these studies is directly relevant to the production of antibody by an otherwise poor antigen. The essence is probably that for any given antigen molecule taken into a cell there is always a probability less than 1 of its being incorporated into a template; for some the probability may be very small indeed. Any manipulation by which the number of opportunities is increased will to that extent increase the likelihood of detectable antibody production if the substance has any antigenic potentialities whatever.

(7), (8) Embryonic animals do not produce antibody because it is a physiological necessity that in this period an inhibition against auto-antibody production should be built up. The period at which the possibility of active antibody production develops appears to vary considerably from species to species. Cross skin-grafting between sheep fœtuses *in utero* may give typical intolerant reactions, before birth (Schinckel and Ferguson, 1953) and there is some evidence that guinea-pigs injected *in utero* can produce antibody against influenza virus (Dettwiler *et al.* 1940). On the other hand Hanan and Oyama (1954), by repeated injections of bovine serum albumen in rabbits from birth onwards, produced a lasting inhibition of antibody formation against the homologous antigen. Kerr and Robertson (1954) obtained somewhat similar results in showing that calves inoculated with large doses of a trichomonas antigen in the first week of life became non-responsive to subsequent injections. They point out that when access to colostrum is prevented the γ-globulin fraction of serum does not make its appearance until the third week of life, and believe that it is in this period that the acquired immunological tolerance can be induced.

(9) From the data of Taliaferro *et al.* (1952) X-irradiation interferes drastically with some essential early stage in the process. It is well known that X-rays have a highly damaging effect on DNA function and there is a general belief that most of the biological effects of X-irradiation are secondary to its action on DNA. RNA turnover and protein synthesis are not much affected, neither is the production of influenza virus in the de-embryonated egg (Stevens, 1954). It is natural, there-

fore, to look for an influence on some nuclear process as the cause of the inhibition of antibody production. An important point to be kept in mind is that the action on a secondary response has the same general character as on the primary antigenic response. The crucial radio-sensitive phase is evidently concerned in both types of response.

The lymphocytic series of cells is the most radio-sensitive of all types in the body and it is natural to think, therefore, that some process in the stem cells of the series may represent the crucial point. Since on our hypothesis the only way by which antibody production can persist is by the fabrication of protein-synthesizing templates which are genocopies of the primary template, it is logical to assume that the production of genocopies takes place in the nuclei of the stem cells and that this is the radio-sensitive phase.

(10) The difference between primary and secondary response has already been discussed. We do not feel particularly satisfied with the interpretation already given, mainly because it leaves no satisfactory reason for the change in the quality of antibody following repeated immunization. It is well known that in the immunized animal reinjected antigen is rapidly bound by antibody and both antibody and antigen are destroyed, presumably in macrophages for the most part. Iodinated antigens are more extensively broken down than in the normal animal (Laws and Wright, 1952), and it may be that this modification in disposal allows the appearance of new determinants as well as those used in the primary stimulus. We are at a loss to account for the process by which the rapid secondary response is set in motion by entry of a second dose of antigen. The X-ray data suggest that the difference is not so fundamental as we have been inclined to believe. Perhaps it is best to ascribe the greater speed and effectiveness of the secondary response merely to that general character of organisms and tissues by which the effectiveness

of a function improves with performance. There must be more specific reasons, and one possible suggestion which may contain part of the truth is that antibody on the surface of lymphoblasts when it reacts with the corresponding antigen causes local damage, perhaps including histamine liberation and that this acts as a general stimulus to protein synthesis and cellular proliferation.

(11) Although the fact is well established that there are inheritable differences in immunizability, not much evidence is available as to the possible mechanisms. Two interesting examples from human pathology may be mentioned. There are, first, the cases of congenital lack of γ-globulin in children described by Janeway *et al.* (1953), most of whom show failure to develop antibody or immunity and who would presumably never have survived in the days before effective antibacterial therapy. The second is the finding by Kuhns and Pappenheimer (1952) that there are some individuals who produce non-flocculating antitoxin after immunization with diphtheria toxoid. These are persons with a strong history of allergy and in most respects their abnormal antitoxin resembles 'hay-fever' type antibody. This instance may be of importance in indicating that immunological modification can be impressed by one antigen on two different types of protein-synthesizing mechanism. In general, minor biochemical deviations of genetic origin are so common as between individuals, races or species that it would be quite extraordinary if some apparently arbitrary differences in immunizability did not appear.

(12) References to variations in avidity, range of cross reaction and proportion of monovalent antibody obtained from the same antigen according to species, age of animal and course of immunization will be found in Burnet and Fenner (1949). This again is biologically inevitable where we are dealing with complex patterns carried by molecules which,

though functionally equivalent, are by no means necessarily identical in detail. No biochemist would claim that a presumptively pure preparation of bovine γ-globulin is exclusively a population of identical molecules. A bacterial vaccine is, of course, vastly more heterogeneous in its molecular content. We assume that only relatively small segments of protein, polysaccharide, etc., are built into the RNA+ templates and it is more than probable that the segments utilized vary with circumstances. There will be similar minor difference amongst the RNA+ templates and both these sources will play their part in producing the relatively heterogeneous population of antibody molecules of varying specificity that are found in every immune serum. It is one of the characteristics of macromolecular patterns that the concept of chemical purity becomes almost meaningless.

(13) Sensitization reactions are dealt with in chapter III, sections 1 and 2.

(14), (15), (16), (17) From the most general point of view the cells responsible for antibody production are only important in so far as their limitation to one system points to the process being a specialized functional activity, not an intrinsic function of all vertebrate cells. As has been indicated already, there is no serious controversy over the conclusion that antibody production is a function of these organs or temporary cell accumulations where fixed macrophages and cells of the lymphocytic series are in intimate relation. The antigen is taken up initially in the macrophages, the antibody is liberated from plasma cells, and rather less certainly, from lymphocytes.

It is still impossible to be dogmatic about the site of the other processes which it seems necessary to postulate on functional grounds. On the very slender evidence of the radio-sensitivity of an early essential phase of the process, and of the fairly regular entry of antigenic material detectable by fluorescent antibody into the nucleus of the reticulo-endothelial cell, we might suggest that a nuclear process is important and that what we regard as the central phase of the process—the production of a genocopy of the primarily modified template—takes place in the nucleus. We do not feel that there is adequate evidence to decide whether the nuclei concerned are those of reticulo-endothelial cells or of the reticulum cells from which plasma cells will eventually be derived.

(18) In view of the analogies that have been drawn with adaptive, enzymes it is probably important to stress that there is no real evidence that antibody acts in any conventional sense as an enzyme for which antigen is the substrate. When the nature of the antigen is appropriate and techniques are available for removing antibody from the ⌜precipitated antigen, antibody⌝ complex, intact antigen can be regained.

(19), (20) The contention that antibody can be produced after all antigen has been eliminated from the body has been the aspect of the Burnet-Fenner theory that has been most unacceptable to a majority of immunologists. A considerable amount of effort has been devoted to an experimental study of the question and we have already reviewed the results of those experiments.

It is possible, however, that the question has never been put in the proper form. Everything suggests that the greater part of the molecule of any antigen plays no part in determining the specificity of the antibody produced. It is well known that antigen is multivalent in respect to antibody and our conception of the average protein antigen would be something like Fig. 5 in which A and B are two short amino acid sequences which are the actual antigenic determinants. Whatever form of template hypothesis is adopted it will be the groups A and B, not the whole molecule, which supply the specificity.

The real alternatives towards the solution of which experimental effort should be directed can be stated as follows: (i) is the template responsible for conferring specificity on antibody always one in which the actual atoms of the foreign determinant groups have persisted? or (ii) is the function of the foreign determinants merely to allow a master template to be laid down from which, by direct or indirect replication, an unlimited supply of other templates can be produced? In our phrase these are 'genocopies' of the 'master template'.

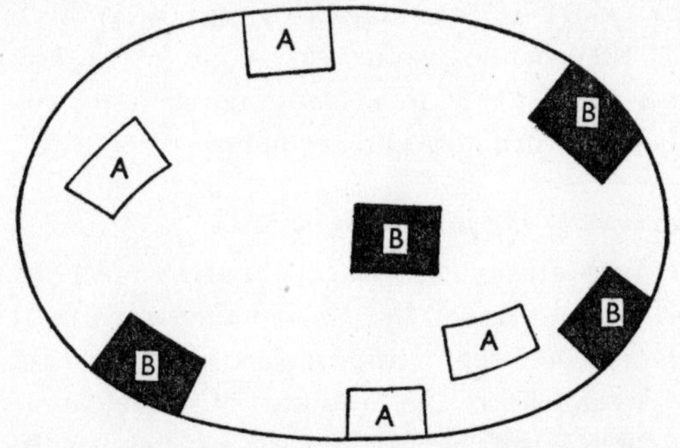

FIG. 5. A diagram to illustrate the possible distribution of two antigenic determinant groups, A and B, on a protein molecule.

On any form of the Haurowitz, Mudd, Pauling hypothesis all the templates are originals, there are no genocopies.

The type of experiment needed would be the labelling of the determinant, and only the determinant, groups of a suitable antigen. Of the systems that have been reported, the most hopeful seems to be that of Clinton *et al.* in which gelatin, itself non-antigenic, was rendered antigenic by the attachment of tyrosine glucoside units. If one or more of the carbon atoms was heavily labelled with C^{14}, it should be possible to follow the fate of the determinant group. It would be essential to get a 'good' antibody response and to show that the standard relation between primary and secondary response holds. A quantitative study of a secondary response along orthodox lines should then settle the question.

One of our main arguments against the view that all antibody molecules are shaped against the template containing the actual antigen is the difficulty of finding where such templates can be stored in the rapidly changing populations of the cells concerned. Where, for instance, do the colonies of obviously newly formed immature plasma cells that Coons finds in the medullary cords of lymph nodes receive their templates from? If templates can replicate, the answer is simple. If they cannot, some very complex *ad hoc* assumptions need to be made to account particularly for the difference between primary and secondary responses.

5. *Weaknesses of the present hypothesis*

In the 1949 statement (Burnet and Fenner) we tried to state the weaknesses of the theory then presented, and to suggest new experimental approaches. In the meantime, one of these weaknesses, the absence of direct experimental evidence for the self-marker hypothesis, has been remedied.

The advance in the knowledge of adaptive enzymes and, in particular, the evidence that, for the well-studied examples at least, the adaptive enzyme can be produced only in the presence of the substrate, has allowed a more specific statement of a possible mechanism which includes the new genocopy concept. We would agree with any critic who contended that there was no substantial evidence (i) that RNA with or without accessory material carried the template for protein synthesis, or (ii) that an antigenic determinant, by becoming attached to the RNA+ template, would directly impress a complementary pattern on the protein being synthesized. Both are merely hypotheses which, in the present state of knowledge, seem helpful ways of co-ordinating the facts. The vital feature of the concept is expressed in the genocopy hypothesis, viz. that the antibody-producing mechanism has a capacity to produce in another medium an effective copy of the essential pattern of foreign antigenic determinants. In our opinion, if that is to be controverted, the experimental approach must be along the lines discussed in the previous section.

In the earlier form of this theory, it was assumed that the effect of antigen was to compel a modification of the recognition unit (RU) of the type most nearly analogous to the antigen. With partial adsorption a situation of strain develops which can be relieved by a process similar to that postulated by some writers for adaptive enzyme formation, i.e. the resynthesis of the RU, presumably by transpeptidation so that an effective complementary fit takes place. This was then followed by a direct or indirect reshaping of the RNA+ template to correspond. This is more elaborate and we should think less likely than the alternative which has been adopted. Its only advantage is that, when essentially the same relation is assumed between secondary and primary stimulus as between primary stimulus and formation of the analogous recognition unit, we have a somewhat more satisfactory approach to the problem of changing antibody quality with repeated immunization.

Possibly the main residual weakness of the present formulation is its reticence about 'immunological paralysis' in the sense used by Felton (1949). The phenomenon is typically seen in mice inoculated with purified pneumococcal polysaccharides. If the dose is a small one of the order of one-tenth of a microgram, mice will be firmly immunized against experimental infection with the corresponding type of virulent pneumococcus. If a larger dose (0·1 to 1 mgm) is given, the mice injected not only fail to become immune, but also lose any capacity for subsequent immunization. For as long as a year afterwards the mice cannot be immunized by a dose of antigen which is highly effective in normal mice of the same age. This prolonged immune paralysis is associated with retention of the polysaccharide in the tissues where it can readily be detected by Coons's fluorescent antibody technique.

A re-examination of this phenomenon by Stark (1955) using pneumococcal polysaccharide labelled with C^{14} leads to a relatively simple explanation in terms of this persistent deposit of antigen. Stark finds that, judged by radioactivity, the material remains in undiminished amount in the tissues. It is apparently not susceptible to the normal processes of metabolism. If, however, the antigenicity of tissue extracts is examined by inoculating them into other mice, a progressive fall is observed. This is ascribed to the antigen deposit becoming progressively blocked with adsorbed antibody. In favour of this view is the fact that a mouse immunized by a small dose of polysaccharide will lose its immunity if subsequently inoculated with a large (paralytic) dose.

A few other phenomena of somewhat similar character are mentioned in the literature. By 'saturating' a rabbit with daily injections of a soluble foreign protein like bovine serum albumin over a period of 6 to 12 weeks, it is possible to produce a virtually complete unresponsiveness (Dixon and Maurer, 1955). With adult rabbits it is not possible to retain this unresponsiveness after the animals have been rested. With animals in which the saturating injections were commenced soon after birth, their specific unresponsiveness could continue after all antigen had gone from their tissues. Hanan and Oyama (1954) had earlier obtained similar results in rabbits inoculated from birth with bovine serum albumin given intravenously. After three injections weekly for 15 weeks and a few weeks' rest, these animals gave no antibody response to the homologous antigen. Some also failed to respond to a heterologous antigen egg albumin. Taliaferro *et al.* (1951) found a similar temporary failure of antibody (haemolysin) production in rabbits given large amounts of foreign red cells in multiple intravenous injections. Kerr and Robertson (1954) found that if very young calves were inoculated with large doses of *Trichomonas fœtus* antigen there was no antibody response and when the calves were retested some months later there was again little or no antibody production. The effect was dependent on the size of the dose, small initial inocula producing only partial impairment of subsequent responses.

The impression one receives is that the production of recognition units in embryonic life may not be so clearly differentiable from antibody production in the developed individual as we have tended to assume. It will be remembered, however, that in fact the essential difference postulated between an antibody unit and a recognition unit concerned only the nature of the secondary processes. Renewed contact with the antigen stimulated the first but not the second to the characteristic activity of protein synthesis and cellular proliferation. There are interesting resemblances which may be more than superficial between the cellular changes associated with active antibody production and those seen in some types of virus infection. In a sense we can regard the antibody unit that is transferred to the lymphoid stem cell as a virus-like agent capable both of replication and of stimulating the cells it infects to proliferative activity. There may be much to be learnt from a comparison of antibody production with some of the aspects of influenza-virus multiplication in the host cell. In later sections the phenomena of interference and incompleteness in influenza virus are discussed at length. The interpretation of these phenomena gives some basis for the suggestion that the behaviour of a potential antibody-producing cell overloaded with antigen may be analogous to that of a susceptible cell infected simultaneously with several virus particles. In neither case can the normal product be liberated in fully active form.

The type of antibody produced as a result of secondary processes varies according to circumstances even with the same antigen. It is probable in fact that by choosing appropriate conditions a single antigen could be used to induce production of either (i) classical circulating antibody, (ii) hay-fever type antibody, (iii) tuberculin type sensitization, or (iv) specific immunological tolerance. It is a task of the future to define the appropriate conditions more closely, but it is not unreasonable to believe that specific tolerance (above, (iv)) normally characteristic of the prenatal reaction, may eventually be proved inducible also in cells exposed to concentrations of antigen which are beyond their ability to handle. The possibility that potentially antigenic substances taken in by the intestinal route are specially likely to be dealt with in this way is discussed in a later section.

CHAPTER III

THE SELF-MARKER HYPOTHESIS IN RELATION TO CELLULAR PROLIFERATION AND CONTROL

1. *Immunological aspects of tumour transplantation*

The significance of genetic and immunological factors in limiting the transplantation of malignant tumours to new hosts has been known for many years (Strong, 1926). Recently, however, there has been considerable interest in the possibility that immunological factors may have a relatively direct bearing on the nature of carcinogenesis as well as on transplantability of developed tumours. There seems to be no doubt that with the development of malignancy and experimental transplantation, cells can be obtained which will grow progressively in hosts that would have reacted strongly against the normal cells from which the transplantable tumour was derived. This presumably results in part or wholly from loss of antigenic specificities—'self-markers' in our terminology—from the cells in question. Green (1954) has recently developed concepts of the nature of the cancerous cell in which loss of self-markers is regarded as a central feature in the development of malignancy rather than a more or less regular concomitant of the process. His ideas arose as a result of studies on the tumour-inhibiting action of carcinogens, a phenomenon well known from the work of Haddow and others. He found that the tumour-inhibiting quality could often be found in tar fractions free of known carcinogens; and several pure compounds, including 3:4-benztetraphene, with tumour-inhibiting power were isolated. These are non-toxic, do not inhibit somatic growth and are not carcinogenic. The tumour-inhibiting effect of benztetraphene was best shown with transplantable tumours of the rat, and its effect was greatest the earlier it was given. With transplantable tumours propagated in pure line strains of mice the effect of benztetraphene was much less. The impression was strong that the action of benztetraphene was to strengthen the normal immune response against the transplantable tumour. Where there was no possibility of response there was no action of benztetraphene, e.g. it was inert if cortisone was given to the animals. It seemed, therefore, that benztetraphene acted by combining with protein or other cell component to produce the same antigen which, in the transplanted tumour cell, provoked the body to an immune response which might or might not be effective.

Another indication of the involvement of immune processes was seen in the marked accumulation of mature and immature plasma cells in related lymph nodes and, at the time of regression, in the substance of the tumour.

From these and other findings, Green postulates that the essential change associated with malignancy is a loss of identity antigens which he considers are probably to be equated with the self-markers of Burnet and Fenner's (1949) theory. In discussing the problem it will be more convenient for us to use the terminology already adopted rather than to use Green's terminology, and to look first at the orthodox immunological interpretation of resistance to transplanted tumours in animals which are not genetically identical with the donor of the tumour.

When a transplantable sarcoma retrogresses in a rat something like the following process presumably takes place. The transplantable cells liberate a (series of) protein antigen(s) foreign to the body which passes to the local lymph node and to other antibody-producing centres. Antibody is produced, probably of the sensitizing type, and largely carried by lymphoid and plasmacytoid cells. Antibody and antibody-carrying and sensitized cells are attracted to and react with the tumour region, causing regression in much the same fashion as occurs with normal tissue homografts. The postulated effect of benztetraphene and other tumour-inhibiting hydrocarbons then is to combine with local cells or cell products to produce material essentially foreign which can provoke the same type of (sensitizing) antibody as is provoked by the foreign tumour cell.

Green's view of carcinogenesis, e.g. by 1:2:5:6-dibenzanthracene, is that in the first step dibenzanthracene combines with a marker constituent of normal epithelial cells. This will result either directly or indirectly in a modification of the specific marker pattern so that when those marker molecules that are liberated from the cells pass to the local lymph nodes, they will provoke an antibody response. This will presumably be of the type concerned in homograft reactions and will tend to handicap cells that are producing the modified antigen.

If application of the carcinogen is continued, there will be local irritation accentuated by immune response, histamine liberation, etc. This irritation will allow increased proliferation of epidermal cells which, however, will be at a disadvantage because of their saturation with the artificial compound antigen which has provoked the active antibody response. Should a somatic mutation occur by which a self-marker is lost, then either one possible point of attachment or the *only* point of attachment for carcinogenic and tumour-inhibiting hydrocarbons will have been lost. This will immediately provide an advantage to the cell and, if the conditions are appropriate, it will proliferate more effectively than its unmutated congeners. Green suggests that it is the loss of self-marker components that in itself makes a cell malignant.

While this may be true, it cannot be accounted for purely on immunological grounds. Once a carcinoma has been induced it keeps on growing despite cessation of the application of the carcinogen. By hypothesis there are now no foreign markers being produced and at least one of the normal self-markers has been lost. Circulating antibody produced by the carcinogen-modified markers would now be expected to have no more influence on the carcinoma cells than on normal ones. If loss of self-markers by somatic mutation gives rise to malignancy, then it becomes necessary to look for other possible functions of the self-markers that, from Medawar's experiments, we can be sure are being produced in and liberated by skin epithelium. On the reasoning so far used it is not immediately clear why such cells should have markers at all. They are expendable, but they are not dealt with by the reticulo-endothelial system. They are progressively pushed outward by subjacent growth, become more and more cornified and are eventually discarded. Since under normal conditions worn-out skin epithelium will never be taken up by scavenger cells, the significance of self-marker production must be wider than the one we have so far ascribed to it.

In an attempt to clarify possible relationships between the self-markers of immunological theory and the phenomena of carcinogenesis and tumour transplantation something may be gained by examining the implications of the hypothesis—that an essential part in the control of cellular relationships (morphology) within the body is provided by the liberation of marker molecules by each functioning cell. These may function either by influencing other cells in immediate contact with the liberating cell or by passing in lymph and blood streams to other sites.

We may consider first skin epidermis as the tissue favoured for transplantation experiments and the site of carcinogenesis in many experimental cancer studies. Here we have a tissue composed of epithelium under constant and controlled proliferation. The cells are expendable, becoming, as they are pushed upward, gradually cornified and are eventually discarded to the outer environment. They are never under normal circumstances got rid of by the scavenger reticulo-endothelial cells. Yet they clearly have antigenic qualities in common with expendable cells of normal type such as those of the blood. Medawar's group have shown that tolerance to a foreign skin graft can be induced by inoculation of blood of the corresponding type into the fœtus, the test being made when the mice treated *in utero* are a few weeks old. Any self-markers that are concerned in determining the disposal of red cells then must also be present in skin epidermal cells. This is in line too with the fact that in many cases resistance to an otherwise transplantable tumour can be induced or increased by inoculation of blood cells from normal animals of the strain which serves as donor of the tumour (Barrett and Hansen, 1953).

Further important information is to be gained from the behaviour of a tolerated skin graft of strain B in an A-strain mouse that has been prepared by fœtal inoculation of B cells. If in a mouse in which the B graft is fully tolerated a normal lymph node from an untreated A mouse is grafted, there develops after a few weeks an intolerance of the graft which is destroyed and got rid of in much the same way as occurs with a primary graft of heterologous skin in a normal recipient (Billingham *et al.* 1955). We can hardly avoid postulating, therefore, a steady liberation of molecules bearing the immunological imprint of the tissue concerned—a self-marker, in other words. This self-marker (we may use the single form for simplicity until it is shown that more than one type is required) is the same as is produced by, or present in, a variety of other cells including some present in kidney, spleen and blood.

The implications of Anderson *et al.*'s (1951) findings that dizygotic twins in cattle will readily accept cross transplants of skin may also be important. Cattle that are full siblings but born from different pregnancies are as intolerant of mutual transplants as unselected cattle of similar breed. Here again we have evidence that essential marker molecules from both individuals pass to all potential antibody-producing centres and provoke an acquired specific tolerance towards all cells or, at least, towards the liberated markers of all cells of both genetic types. This suggests rather strongly that the genetically determined quality of a cell which governs its acceptability on contact with other cells of the body is mediated through diffusible markers.

At the genetic level there is also much of significance in the fact that the F_1 generation from two homozygous stocks of mice will accept transplants from either parental type. The parent stocks, however, will not accept transplants from each other or from the F_1. This is one of the instances in which the relation between gene and antigen appears to be particularly direct. If we take any gene concerned with cell specificity differences A and a, a cross of $AA \times aa$ will give F_1 all Aa. If A and a each give rise to corresponding diffusible markers, the results are as would be expected. If, however, self-markers arose by a secondary process involving interaction amongst differentiating cells and components, one would not expect the simple relation to hold. This leads to the tentative conclusion that most markers from all types of cell have a common character determined directly by the genotype. It is, of course, not incompatible with this that there should be a number of other potential self-markers in or on expendable cells such as erythrocytes. These may well be of different quality from the diffusible tissue-cell markers with which we are at present concerned.

When cells with the 'wrong' marker appear in the body this marker will, by the same mechanism, pass to the lymph nodes and provoke antibody formation (probably of sensitization type) and therefore an inflammatory response in the region from which the marker antigen is diffusing. This will result eventually in elimination of those cells which liberate the marker.

The same or similar behaviour will presumably be shown by other tissues than the skin, and most workers would include as well all potentially transplantable tumours induced or arising spontaneously in the animals concerned. It is the general rule (Snell, 1953) that a malignant tumour arising in a pure line stock of mice (A) can be transplanted to any other individual of that stock but not to animals of a genetically distinct strain (B). It will grow in all F_1 hybrids of A and B but in the F_2 generation will grow only in a proportion of animals. The larger the number of dominant genes which must be common to donor and recipient to allow transplantation, the smaller the proportion of susceptible F_2 individuals. Snell considers that the highest number of histocompatibility gene differences that has been recognized between two strains of mice is 14. By implication a similar number of differences of one sort or another must exist between marker molecules (or structures) of the two types.

Strong (1926) and many subsequent workers have shown that with repeated transplantation a tumour becomes less specific in its requirements. One which originally required the correspondence of 8 genes to allow effective transfer gave rise to lines needing only 2 or 1 such correspondence. In the process the tumour cells must have lost self-markers or eliminated some of their specific pattern.

A striking recent example is given by Koprowski (1955). He used an ascites tumour specific for C_3H mice and rendered mice of another strain susceptible by injecting fœtuses *in utero* either with C_3H blood or tumour cell emulsions. In this way a line of tumour cells was developed which could be transferred successfully to any strain of mice. Koprowski suggests that the experimental conditions provide especially suitable conditions for the selective survival of mutant cells which have lost some of their marker characters.

A diagrammatic indication of this might be attempted by the formulation indicated below. Here **1** and **2** represent closely related stocks of mice of one strain and **3** and **4** stocks of a second distinct strain.

1	2	3	4
A	a	α^1	α^2
B B	B B	β β	β β
C C	C C	C C	C C
D D	D D	D D	D D

Family differences are shown only at the top level of symbols A, α, etc., strain differences at the second level B or β.

On this view, if **1** was the marker for cells of a mouse in which some were becoming malignant, loss of A would allow transplantation to **2**, but loss of both A and B would be needed to allow transplantation to mice of types **3** and **4**.

Only a minor elaboration of this formulation is needed to provide a picture of Green's phenomenon in which treatment with a tumour-inhibiting non-carcinogenic hydrocarbon prevents the 'take' of a not too well adapted transplantable tumour. We will probably have to assume that both carcinogen and inhibitory hydrocarbon, by combining with the marker, produce a new pattern foreign to all the strains of host under consideration. Not only is A altered, but also B.

Carcinogen	Inhibitor
A++	A+
b b	b b
C C	C C
D D	D D

When by one mechanism or another cancer appears, we have presumably a marker in which the specific component is lost but (? a genocopy of) one of the secondary changes in the marker common to both is retained. If the cancer cell is a very poor liberator of markers while the inhibitor-treated cell liberates the modified marker normally, then the anti-b component should have an inhibitory effect on the establishment of the cancer cell with its

b b
C C
D D

marker, beyond what would be stimulated by this marker itself.

It seems clear that a hypothesis of this type will account for Green's experimental findings and the gradually increasing range of genetic types to which a tumour can be transferred. So far, however, it has no obvious bearing on the essential nature of the cancer cell itself.

2. *The implications of cutaneous sensitization to simple compounds*

Green's hypothesis of carcinogenesis has obvious relationships to the phenomena by which certain small molecules such as chloropicrin can give rise to an immunological sensitization of the skin. The possibility that in these reactions the sensitizing chemical unites with the self-marker of the skin cells should therefore be discussed before attempting to look for any wider significance of the self-marker molecules.

The experimental data indicate that reactions in guinea-pigs are reasonably similar to those observed in human beings and, if we combine clinical and laboratory experience, the essential points that must be kept in mind are as follows:

(1) A wide variety of natural and synthetic substances may be responsible: primula, paraphenylenediamine, 2 : 4-dinitrochlorobenzene, picrylchloride, nickel salts, represent a few examples.

(2) The sensitization is acquired and takes about a week or more to appear after an adequate sensitizing contact or injection.

(3) Sensitization is general to the whole skin. The reactivity is conferred on the skin by some general process not a local one. Haxthausen (1943) described an experiment in which two pairs of identical twins were used. One of each pair was sensitized to 2 : 4-dinitrochlorobenzene and an area of skin interchanged. After healing, tests showed that only the sensitized individual responded; his skin on the non-sensitized twin gave no reaction.

(4) Limited passive transfer is possible in guinea-pigs by the use of washed white cells, probably lymphocytes. Serum is not effective.

(5) In some instances guinea-pigs, which have been fed allergenic chemicals develop a resistance to sensitization by normal skin applications of the allergen. There is no evidence of blocking antibody, and in Chase's view the refractory state is due to an interference with the host's response to a specific antigenic stimulus. Such animals can develop sensitization by transfer of white cells from a sensitized animal (Chase, 1952, Battisto and Chase, 1955).

(6) There is much to suggest that tuberculin type sensitization and the reaction against foreign tissue transplants are basically similar to these contact sensitizations.

These results and deductions bring the phenomena into line with classical antibody reactions if one assumes that union of the chemical with host protein produces an antigen of the same general quality as a foreign marker. The antibody produced in such cases is not detectable in serum, but is widely diffused through the body by one of two methods: (i) by transport on or in cells, lymphocytes(?), with transfer to capillary endothelium in appropriate situations; (ii) by transport in soluble form in the blood but with so rapid a removal that at any given time only traces are present, the bulk being rapidly taken up by the vascular endothelium.

Although little evidence is available on the point, one would assume, by analogy with the behaviour of classical antibody and from Haxthausen's experiment, that antibody fixed to the tissue endothelium would have a half life of days or at most a few weeks.

The most interesting aspect from our present point of view is the indication that sensitization of this type is more or less specific for the skin. This may be very largely due to concentration of clinical and experimental interest on easily visible reactions, but at least it appears to be well established that guinea-pigs cannot be sensitized to agents of this type by intravenous or intraperitoneal injection (Simon *et al.* 1934). Further, Chase's results by feeding add the suggestion that by other routes the chemical agent has an effect on the antibody-producing centres almost analogous to the tolerance produced by prenatal exposure.

There is general agreement that the sensitizing antigen is a complex of the small molecule with a skin protein and it is economical of hypothesis to regard the skin markers as the most likely host component to be involved. This is in line with the type of immune response produced and with the fact that marker molecules are constantly passing to the lymph nodes. There is relatively little information on whether sensitization reactions of this type can be elicited in other tissues than the skin. Gell (1944) showed that after intracutaneous injection of the sensitizing chemical 'tetryl' into guinea-pigs anaphylactic reactivity of the gut or uterine muscle could be demonstrated as well as the skin response. Sensitization could also be transferred passively by serum. This is never observed with contact sensitization of the skin. When artificial conjugates of the sensitizing chemical with a foreign protein are used for immunization, it is usual to find that precipitating antibody of classical type only is produced. Although the specificity is largely determined by the nature of the sensitizing substance used, skin sensitization is not produced. This was shown by Gell *et al.* (1946) in rabbits and by Chase (1955) in guinea-pigs. In the guinea-pig experiments it was found in addition that lymphocytes from such animals could transfer antibody-producing capacity to other guinea-pigs but did not sensitize them.

The significance of sensitization for antibody theory has never been adequately assessed and it is doubtful if there is adequate factual knowledge to justify such an attempt. In our 1949 monograph, the differences between classical antibody, 'hay-fever' type, antibody and 'tuberculin' type antibody were regarded as associated with the firmness of cell-antibody association. If we call these types 1, 2 and 3 respectively, we have a progressive increase in the ratio (antibody bound to cells)/(antibody in body fluids). Cole and Favour (1955) have recently given the first satisfactory evidence that tuberculin-type antibody can be present in serum from sensitized animals. Using large amounts of sera from tuberculous guinea-pigs they obtained by Cohn's methods a fraction IV 10, which could passively transfer delayed-type tuberculin sensitivity. It seems still possible however, that type-3 antibody is largely transported in cells and transferred to vascular endothelium by direct cellular contact. It may be recalled here that tuberculin-sensitive cells in tissue culture give rise to tuberculin-sensitive progeny. This was interpreted by Burnet and Fenner as indicative of replication of the antibody-producing mechanism in the cells concerned. All this suggests that it would only be a slight further step for entry of the sensitizing antigen to provoke a tolerance equivalent to the prenatal type, which might perhaps be called type-4 antibody.

There is almost certainly something of significance for

general immunological theory in this sequence. There are three points that may eventually prove relevant for its interpretation.

(1) There is in general a diminution in the size of the 'antigen' molecule as we pass downward; the active agents of ragweed pollen and of tuberculin, both seem to be relatively small polypeptides which may need to combine with host constituents to become complete antigens.

(2) It might be expected on general evolutionary grounds that potential antigens taken in by the intestinal tract would be rendered inert by one mechanism or another in all normal individuals. There may be some relevance of Chase's experiments in this connexion.

(3) Hay-fever type antibody (type 2) is provoked predominantly by respiratory inhalation with presumed passage of antigen through the respiratory mucosa. Tuberculin (type 3) sensitization is associated with the presence of the antigen in a subacute inflammatory mass in the stimulation of which waxy substances may play a significant part (Raffel and Forney, 1948). Contact sensitization (type 3) is a function of the skin. Chase's specific induction of tolerance to sensitizers (type 4) is via the intestinal tract.

Rather vaguely one pictures the sequence of effective antigens as comprising in type 1 no host component, but in the others an increasing ratio of host protein to foreign determinant from types 2–4. It is tempting to equate this characteristic with the capacity of the corresponding antibody to associate with tissue cells, but without much more factual data no elaboration of the idea would be justified.

In summary the general impression that emerges is that there may well be significant resemblances and analogies between the phenomena of skin sensitization and chemical carcinogenesis. Advance beyond this will probably only become possible when the self-marker concept is clarified.

At present it is purely a functional concept which would be sharpened greatly by a knowledge of the type of macromolecule involved as carrier of the pattern. The way in which differences, known from immunological and genetic results to exist, are expressed in these macromolecules is quite unknown. The various marker qualities may be inherent in every protein molecule produced in the cell or they may be confined to one or more specific molecular species.

3. *Application of Weiss's concepts of cell control to the self-marker hypothesis*

In an attempt to press the self-marker idea a little further in relation to the problems of the control of cellular proliferation, we may look more closely at the types of 'expendable' cell found in the mammalian body. Three types can be recognized: (i) red cells and lymphocytes which are taken up by reticulo-endothelial cells and disposed of; (ii) endocrine cells such as those of thyroid and pituitary which are controlled by the level of circulating hormones; (iii) expendable epithelium such as that of skin and intestine where effete cells are liberated into the environment.

Experience with tissue cultures and in many fields of experimental pathology suggests that in the presence of the needed food supply, living cells will continue to proliferate until they are restrained by some inhibitory agent or situation. A morphologically defined tissue must, therefore, be made up of cells whose intrinsic tendency to indefinite proliferation is effectively controlled. Simple contact with other cells may be a potent inhibitory factor, but on general grounds one would feel certain that 'simple contact' is in fact a very complex and subtle interaction between the two surfaces. Structural control must be, to some extent at least, homœostatic automatic control, in the sense that proliferation should give rise to a condition that inhibits further proliferation and that the appearance of need for proliferation should be associated with the opposite phase of the condition.

Nervous control can be eliminated by the fact that denervated regions show no gross or direct breakdown of morphological control. Virtually the only possibility left is that control results from an interchange of significantly patterned macromolecules or possibly a unidirectional control of this type. In the case of the thyroid, when the specific product of the gland is at a low level in the blood, this low level removes a restraint and the pituitary secretes a larger amount of thyroid-stimulating hormone which provokes proliferative activity in the thyroid.

If we look at the simpler examples of cellular co-ordination as in the skin epithelium, a possible analogy to the endocrine situation would be to find that an adequate concentration of marker molecules from adjacent cells is a necessary requirement to keep proliferation in check. The marker molecules drained off to the lymph nodes to function, when abnormal, as antigens would then be a necessary by-product of this function.

To carry this hypothesis further we must postulate that there is an exchange of marker molecules (SM) between any two adjacent or related cells, and as a corollary, if they are to have any function, that there is a recognition unit (RU) by which the 'instruction' can be accepted. One would naturally imagine that relationship between the RU and the self-marker (SM) has been developed out of the processes by which the two agents are synthesized, but for the time being this may be neglected and the two entities SM and RU accepted as given. The hypothesis then is that the cells are kept from proliferation by the existence of a dynamic interchange of recognizable particles which means, in fact, the maintenance of an adequate concentration at the interfacial area and a constant drain of particles not 'recognized' to the lymphatic system.

This hypothesis of self-marker and recognition units, although reached from a different point of view, is basically very close to the ideas on morphogenesis that have been developed by Weiss (1947, 1950). He considers that a cell will respond to substances present on adjacent surfaces of other cells by the sorting-out of substances from its metabolic pool which have a specific complementary pattern (SCP) relationship to the 'adjacent' substances. Where the two cells are of the same type, Weiss considers it probable that there will be a symmetrical relationship, in that cell 1 will produce units A which will provoke the organization of complementary units *a* on cell 2. This will also produce A units which will provoke *a* on cell 1. Owing to the existence of such mutually complementary patterns, two similar cells will tend to remain attached and proliferative activity will be inhibited. Except for assuming a more dynamic type of interaction—itself very much in line with Weiss's concepts of molecular ecology—there is little formal difference between the two points of view.

There are several possibilities when by one means or another abnormal self-markers are produced by a cell, either because the cell has been transferred from another host and is genetically different or by the action of a chemical. One can imagine at least two grades of difference. The first type of abnormal marker is not recognized as different by adjacent cells but is by the specialized cells of the scavenging, antibody-producing mechanism. Here we can take as examples skin cells transplanted from another homozygous strain of mice or modified *in situ* as the result of treatment with sensitizing substances of small molecular weight.

The second type of difference might be so extensive that the self-marker could not carry out its proliferation-inhibiting function. In our view some such concept is implicit in Green's immunological view of carcinogenesis. Further examination of Green's hypothesis must be based on the results of experimental studies on the chemical induction of cancer. A limited survey of current reviews, notably Wolf (1952) and Haddow (1953), suggests that the following points are relevant.

(1) Initiating and promoting action are to be differentiated. The first stage is specifically carcinogenic, taking place very rapidly, essentially irreversible and producing a number of latent cancer cells proportional to the dosage of carcinogen; the second stage is produced by irritants such as croton oil which develop the latent cells into overt malignancy (Berenblum and Shubik, 1947 *a*, *b*).

(2) The growth-inhibiting effect of carcinogenic hydrocarbons was found by Haddow and is the basis of his general theory of cancer (Haddow, 1938), which is approximately that the carcinogen has an inhibitory effect on cell growth providing an unfavourable environment, to overcome which an adaptive and irreversible change—somatic mutation—occurs, giving rise to a new race of (cancer) cells.

(3) A different type of agent, substituted azobenzenes, is concerned with liver cancer, but only functions under conditions of dietary deficiency. The aromatic amines, e.g. acetylaminofluorene, have a somewhat different distribution of carcinogenicity also (Wolf, 1952).

(4) Para-dimethylamino-azobenzene ('butter yellow') combines directly with a liver protein only in those animals and tissues susceptible to its carcinogenic action—the compound is probably broken down as formed and eventually exhausts the power of the cell to go on producing. Then some may fail to synthesize the protein at all and, in the process, take on malignancy of growth (Miller and Miller, 1948, 1952).

(5) Green has established that some inhibitory hydrocarbons are not carcinogenic.

(6) Sustained demand on endocrine cells (or sustained stimulation) will cause hyperplasia, adenoma and carcinoma. Thiouracil blocks the capacity of thyroid cells to synthesize thyroxin; low thyroxin increases the pituitary output of thyroid-stimulating hormone with development of hyperplasia proceeding to carcinoma in the thyroid (Purves and Griesbach, 1946). Similarly, the destruction of thyroid by I^{131} causes pituitary tumours, the absence of thyroxin having removed the normal control. (Gorbman, 1949).

(7) Prolonged tissue culture (Earle *et al.* 1943) of a pure cell race can in itself induce a malignant change.

The differentiation of initiating and promoting actions may be associated with the fact that a single malignant cell which has, by a somatic mutation, lost its capacity to produce normal markers may still be controlled by adjacent normal cells until circumstances make it possible to disturb the situation by irritation with croton oil, traumatization, etc.

The mechanism of the somatic mutation is the most difficult feature of the concept—especially if one has to postulate a genetic change in nuclear DNA. If, however, the self-marker and recognition unit system is laid down in embryonic life by cytoplasmic mechanisms, as has been demonstrated for the RU's associated with the scavenging system, it may be worth trying to develop hypotheses as to possible mechanisms (especially by the use of the genocopy concept). We have a cell producing self-markers from a RNA+ template, laid down originally by genetic processes. These may differ significantly from one type of cell to another and we can imagine that by specialization of those differences we reach agents like the anterior pituitary hormones. As a cell finds itself in relation to others it will, by the mechanism previously postulated for reticulo-endothelial cells, develop RU's to protect itself from any further interference with slightly alien patterns. A cell may, therefore, be expected to have a set of

RU's for its own self-markers—these might well be part of the general protein synthesizing mechanism of the cell—and more specialized ones for those other self-markers which reach it by any route. This would lead naturally to a specialized development of the RU system in the reticulo-endothelial cells. Stability of a cell in any particular tissue is associated with the normal receipt of the various types of markers at the appropriate surfaces of the cell. A carcinogen by hypothesis is a chemical configuration which can be incorporated into a self-marker, so changing its pattern. At a certain concentration we shall find a large proportion of these persistently marked in the new pathological fashion. This will cause a complex disturbance intracellularly and immunologically.

The mechanism of liver carcinogenesis by dimethylaminoazobenzene is known in sufficient detail to provide a number of suggestions along these lines. Miller and Miller (1948) found that the concentration of dye bound by protein in the liver was proportional to the number of tumours subsequently arising in the organ. The maximal level was reached in a few weeks and then diminished despite continued feeding of the dye. The tumours themselves contained no dye bound to protein. Miller and Miller (1952) consider that combination of dye with a specific liver protein is the first step in carcinogenesis; with continued administration of dye there is a drain on the cell to continue producing specific protein and perhaps to provide enzymes to destroy the abnormal compound. A certain stage may be reached when the cells in order to survive must lose the capacity to produce the specific protein by somatic mutation. Weiler (1952) finds that there is in fact a definite loss of serological specificity in butter yellow hepatomas as compared to normal liver. The suggestion is obvious that a self-marker has been discarded. It is unlikely that a change of this sort can be referred back to the

DNA nuclear mechanism on present theories, but there is nothing against an appropriate change in the nuclear RNA which seems to represent the first intermediate station in the chain of command. This would then represent a full somatic mutation in the sense that all descendants of the cell would fail to produce normal markers. If, as seems possible, failure of self-marker production is also associated automatically with absence of the corresponding recognition units, such a cell will be functionally isolated from its surrounding cells. As far as one type of control is concerned, the abnormal cell is in a vacuum and 'at liberty' to proliferate. In fact, one would guess that there are second-grade controls as well and that the cell is in the state of initiated but latent malignancy awaiting an appropriate promoting stimulus to waken it to unrestrained proliferation. The possibility that a sequence of such mutations is necessary should perhaps be considered, although the work of Berenblum and Shubik (1947 *a*, *b*) suggests that the initiating change can be produced very rapidly by a single application of a potent carcinogen. This would hardly be compatible with a series of mutations.

Such a view is rather closely allied to that expressed in Haddow's (1953) recent restatement of his earlier hypothesis. He considers that the primary step in carcinogenesis may be inhibition of certain fundamental processes of genetic synthesis followed by the generation of a new self-duplicating template, chemically and hence genetically modified.

To account for the variations in transplantability of induced tumours, it is clearly evident that we cannot equate loss of all marker characters with malignancy. There must remain qualities which may differentiate tumour cells initiated in a host of strain A from the body cells of strain B so that a transplant to B will retrogress for immunological reasons. This may mean either that there is a hierarchy of different marker molecules or, more probably, a hierarchy of levels in a single marker unit (see p. 88). If the somatic mutation results in loss of the A,a configuration, this could destroy its marker function in the primary host yet allow it to stimulate immune resistance in another strain. Continued transplantation, sometimes successful, in animals giving a poor immune response would, of course, favour the occurrence of further somatic mutation involving the deeper aspects of the marker specificity and such a process would result in a progressively increasing incidence of successful transfers. It is hard to believe that any other quality can be as important as this in determining the transplantability of a tumour.

This section on the relation of the self-marker concept to cancer research is in a sense a diversion from the main theme of macromolecular pattern. Its only real significance is as providing a clear indication that the self-markers postulated on purely immunological grounds probably have a wider significance. In Chapter 2 the difference between self and not-self was developed as it concerned the contrast between the organism as a whole and other organisms which might enter its cells or tissues from the environment. Here we seem to be approaching the logical extension that within the body itself there are differences necessary between self and not-self, in the sense that every cell has a defined relation to other cells of the body and that the mechanisms for defining such relationships may have been adapted in the course of evolution to the specialized immunological functions that we have discussed.

In a discussion of the evolution of infection and immunity it has been suggested (Burnet, 1952) that the basis of immunity must be sought as far back as that first manifestation of the difference between self and not-self by which one unicellular organism can engulf and digest another. In seeking a possible basis for the evolution of the self-marker system one would have to advance, however, at least to the stage where a multicellular organism of definite morphology exists—the conventional Hydra replacing the conventional Amœba, shall we say. From our point of view we must accept as given the processes by which the organism grows to its characteristic morphology. What may concern us are the recurrent emergencies that must be overcome—the necessities to repair traumatic damage and to digest or overcome in some other way foreign material entering the tissues, including potentially pathogenic micro-organisms.

In discussing repair we can best make use of Weiss's concepts already discussed. Across each plane of contact between two cells there will be diffusion of small molecules and slower interactions of large ones. It will be one of the conditions of stability that this dynamic interchange shall go on and be 'recognized' as such by the cells concerned. Weiss pictures this as being mediated essentially by SCP relationships of macromolecules on either side of the intercellular boundary. The situation, however, is probably less static than Weiss's diagrams indicate. It is probably the failure to receive recognizable material at a given surface that 'tells' the cell that the adjacent cell normally present on that side has been removed by trauma or, for some other reason, is not functioning. This is a piece of information, a potential stimulus, and it is sound biology to believe that in response to an effective stimulus an organism will react in that fashion within its physical and functional limitations which will best assure its continuing survival as a species. The obvious response here is for the cell to enlarge and occupy the vacant space, dividing and proliferating if required until the damage is repaired. Again we have to leave unspecified the nature of the morphogenetic field that must be supposed to control the process, except for one point. When the enlarging or proliferating cells come into contact with normally functioning cells, growth in that direction ceases. The 'instruction' to cease growth seems to come from the flow of material from the cell encountered, just as failure to experience the flow is the instruction to enlarge and proliferate.

If we are to bring infection into this picture it must be by way of intracellular digestion. Any phagocytic cell such as the entoderm of a cœlenterate must have a means of reacting to a foreign nutritive particle by rearranging its internal structure in the vicinity of the foreign surface. It may be that the stimulus to do so is not so dissimilar from the stimulus to growth and proliferation. Since the foreign particle does not diffuse 'self' products it could be recognized as a 'hole'. Indigestible foreign particles are characteristically walled off by proliferating cells.

In the absence of any significant body of knowledge about comparative immunology it would be foolish to attempt to understand how the situation as it exists in mammals was evolved. All that is worth noting is that there has always been an intimate association between trauma and infection so that it should not be surprising if repair and immunity have been based on specialized developments from the same basic attribute of cells that must retain an organized relation to one another.

On this view the recognition unit is the specialized representative in the scavenger cells of the universal capacity to 'recognize' the specific products of adjacent cells. It is doubtful whether the flow of soluble markers to lymph nodes and spleen has any specific function. It seems reasonable that the scavenger cells act simply to remove casual protein, etc., from the fluids that reach them, just as they will remove effete cells and cell fragments. It will be characteristic of their situation and function that a wide range of recognition units will be developed and it would not be an unexpected development from the evolutionary point of view for these cells to give rise eventually to the antibody-producing mechanisms with which we have been concerned. The most primitive function is the simple ability of a cell to recognize the presence of a similar cell in contact with it. The commonly quoted example is the way in which, from an emulsion of cells from two different sponges, like cells adhere and come to form small multicellular units which can develop into new organisms. This capacity is refined and complicated to form the basis of morphogenesis, while other potentialities become specialized into the self-marker and scavenger system on the one hand and the capacity for antibody production on the other.

The self-marker hypothesis is obviously only a provisional and rather clumsy makeshift to draw together an important set of immunological and genetic phenomena that are neglected in orthodox immunological theory. Any attempt to follow its logical implications in any detail soon reaches a point where important questions must remain unanswered. This can be illustrated by a problem that has already been touched on in various connexions.

The hypothesis involves the claim that in virtue of the self-markers incorporated in cells or soluble cell products these do not provoke antibody production in their own organism. Everything suggests that even in a single protein there are many hundreds of potential antigenic determinants. There will be many more in a cell. This immediately provokes a fundamental question. Must we assume that every potential antigenic determinant in the body is 'recorded' in each scavenger cell or does the presence of one or a small number of self-markers inhibit the whole process of antibody production? On common-sense grounds the second seems the obvious choice, and it is the one that we have implicitly adopted without looking too closely at its difficulties.

One of these difficulties is in regard to the interpretation of blood group differences in human beings. There is now a large number of serological systems recognized for the human red cell surface, and in addition to differences represented in one or other form in all subjects there are occasional examples of 'private' blood groups in which only a few related persons give cells agglutinated by a certain serum. Omitting the *ABO* groups, all the others have been recognized essentially because, following transfusion or an incompatible pregnancy, a person has developed a serum antibody which will agglutinate red cells from certain people and not those from others. This defines a certain grouping as present on the agglutinated cells and absent on the others. Omitting again most of the finer points of technique and interpretation, in all adequately investigated cases a self-consistent set of genetic relationships can be established. An individual whose blood cells have character *X* never produces antibody against *X*. Superficially this suggests that recognition units for all the potential red cell antigens recognized and unrecognized must be present in every splenic macrophage and presumably in every potential phagocyte in the body. Otherwise the foreign red cells with their alien markers would be expected to provoke antibody production not only against the foreign markers, but also against all those determinants in the foreign cells which were common to the recipient cells.

It would be possible to offer various explanations for the failure of such auto-antibodies to appear. The most plausible would be to point out that all antisera of this type have been automatically cleaned by absorption of antibodies related to the individual's own red cells. Minor production of antibody against a red cell component present in the body's own cells would, therefore, probably produce no recognizable clinical effect and no antibody would appear in the serum. The cells might for a time show a positive Coomb's test or might not.

With such assumptions it would be possible to retain the concept that a very small number of self-markers per cell would be adequate to account for the phenomena. Other difficulties would arise in connexion with transplantation experiments and their interpretation. Would it be necessary, for instance, to provide a marker and a recognition unit for antigens corresponding to all of the 14 histo-compatibility genes in mice that are postulated by Snell? If we adopt the view that the existence of a self-marker can render many potential antigenic determinants inoperative, we immediately meet the question of how close the association between marker and determinants must be to inhibit the activity of the latter. Must they be on the same molecule or is some larger structural unit the basis of the effect ? Other problems arise in regard to the infinite gradations of antigenic relations between the components of different tissues and different species. There is clearly unlimited scope for further experimental study with a reasonable certainty that some more appropriate generalization will emerge to replace the self-marker concept. All that can be claimed for the latter is that it has shown itself to be a stimulating working hypothesis and may well continue for a few years to have heuristic value in more than one biological field. Some of its possible bearings on certain aspects of human medicine were discussed in a recent lecture (Burnet, 1954*b*).

4. *Summary*

Unless one is strong-minded enough to refuse to read any current literature for the whole period involved, it is impossible to maintain a wholly consistent approach throughout a discussion of the present type. At many points it has seemed better to attempt a relatively detailed working-out of hypotheses that are far from being fully established, rather than to keep the discussion at a completely general level. It is inevitable, therefore, that some or all of the details will have to be eliminated or modified as new advances are made in the understanding of protein synthesis in general. It may be wise, therefore, to summarize the two chapters on antibody production in the most general terms in an attempt to differentiate those concepts which are regarded as essential from what is illustrative detail of temporary significance only.

Basically, then, we have attempted to develop *first* a systematic account of the various experimentally established phenomena which are relevant to the biological, rather than the chemical, aspects of immunology. *Secondly*, two principles based on the general concept of specific macromolecular pattern have been suggested as the key to the understanding of antibody production. These have been named the 'self-marker hypothesis' and the 'genocopy concept'. Both are in terms of functional protein pattern and of pattern on any protein-synthesizing template. They are independent of any more detailed knowledge of the mechanism of protein synthesis, but conversely it can be claimed that they present requirements with which any new theory of protein synthesis must comply. The *third* feature of the discussion is an attempt to develop the view that this basic mechanism will respond differently according to different physiological circumstances to give the wide range of phenomena actually observed.

In the most general terms, we can summarize our conclusions as follows:

(1) A pattern on a target substance (antigen) is by a series of recodings, eventually expressed as a specific complementary pattern on host globulin molecules.

(2) In the process the initial use of an antigen pattern as such is replaced by the development of a 'genocopy' of that pattern expressed in the body's own materials—presumably in RNA.

(3) The newly produced, specifically patterned globulin,

either within a reticulo-endothelial cell as recognition unit, or free in the body fluids as antibody, can be regarded as serving a single basic function. By specific union with the corresponding 'self-marker' or antigen, it has what might be called a naturalizing effect which allows the material to be dealt with as if it were homologous to the body.

(4) The replication, transfer and liberation of acquired immunological pattern is governed by physiological considerations which can be related to evolutionary needs.

CHAPTER IV

VIRUS MULTIPLICATION

There is a general feeling amongst virologists that the most general approach to the problem of virus multiplication is to regard it as essentially a taking-over of the host cell's synthetic mechanisms so that, instead of carrying out their physiological function of producing host components, they are diverted to produce components of virus. If this is a legitimate approach, the detailed study of virus infection should throw much light on the normal processes of protein and nucleic acid synthesis.

Unfortunately, really detailed study has only so far been possible with the bacterial viruses. The large particle phages, of which T2 is the type that has been used almost exclusively, are extremely remote in their structure and behaviour from any normal components of plant or animal cells and from plant and animal viruses. The more one appreciates the extraordinary character of these phages, the less one is inclined to believe that their behaviour has any real bearing on other superficially analogous systems. In the past, many people including the writer have called attention to resemblances in the behaviour of phages and influenza viruses. With further study of both systems most of these resemblances seem to have become superficial and inappropriate for serious generalization. The infective particles of the two agents are of the same order of size, but there any resemblance ends. The phage is a completely organized, compact unit with a hexagonal head, tightly packed with DNA, a tubular tail ending in a thickened area of different serological quality and apparently forming a complex adsorptive enzymic structure adapted to allow entry of the virus DNA into the bacterium. The DNA is physically similar to standard DNA from other sources but instead of cytosine it contains a new pyrimidine, 5-hydroxymethyl cytosine not yet obtained from any other source. It contains no RNA nor has there been any suggestion that, apart from traces of bacterial antigens on the surface, any recognizable components with host specificity are incorporated in the infective particle.

By contrast the influenza virus particle is of variable size and morphology and is extremely difficult to differentiate by physical or chemical means from a fragment of host cell cytoplasm. There is no significant content of DNA, and about 0·8–1·0% of RNA in terms of dry weight (Ada and Perry, 1954*b*). The bases in the virus RNA are those normally present in RNA from other sources.

Our main interest in these discussions is to obtain an insight into the processes of protein biosynthesis in mammalian cells. Under the circumstances it seems advisable to eliminate the bacterial viruses from further consideration.

The behaviour of the plant viruses is probably more relevant to our problem. If we confine ourselves to the better studied 'macromolecular' viruses such as tobacco mosaic, bushy stunt and turnip yellows, we find that these are all compounds of ribonucleic acid and protein and that in some, possibly in all, a certain amount of the specific protein of the virus is produced in infected tissues in addition to the complete nucleoprotein. It is unfortunate that up to the time of writing very little is known of the nucleic acid content of the smaller animal viruses. The only analyses reported are those of Eastern equine encephalitis virus which was stated by Taylor *et al.* (1943) to contain only RNA. Leyon (1951) found that purified murine polio virus FA gave an adsorption of ultra-violet light characteristic of nucleic acid. The small polio and Coxsackie viruses are sufficiently uniform in size to give symmetrical two-dimensional crystal-like packing when concentrated suspensions are dried down on the carrier membrane for electron microscopy. In view of the fact that Schaffer and Schwerdt (1955) have made a preliminary statement that human polio virus contains only RNA and protein, one would feel justified in predicting that all the small animal viruses with a diameter about 20–30 mμ and with the capacity for two-dimensional crystal packing will be found to be made up of protein and RNA only.

If this is found to be the case there may be considerable justification for making use of analogies, for instance, from the study of tobacco mosaic virus multiplication in the tobacco plant. It must always be remembered that an animal virus will never convert more than about 0·1% of the protein of the cell into virus protein, whereas up to 80% of the protein in juice extracted from diseased tobacco leaves is actual virus material.

Extracts of diseased tobacco plants, in addition to the virus nucleoprotein, contain significant amounts of an abnormal serologically related protein which has been studied extensively in the last 3 years. Commoner *et al.* (1953) found that the bulk of the free protein was produced late in the process of infection when most of the available nucleic acid had been used up in virus RNA production. Only a relatively small amount of soluble protein with virus specificity is produced; in general only one-twentieth to one-fiftieth of the immunologically reactive material is free of nucleic acid.

Under suitable conditions this protein polymerizes to form short rods which in electron micrographs are indistinguishable from standard nucleic acid containing virus. It is now generally accepted that the nucleic acid in tobacco mosaic virus is situated as a central core of the rod-shaped units (Schramm *et al.* 1955; Franklin, 1955). According to Franklin the protein is arranged round the core in the form of a flat helix. Similar X-ray diffraction studies of the polymerized protein by Rich *et al.* (1955) and by Franklin and Commoner

(1955) are in agreement that the protein forms a shell round an empty (i.e. water-filled) core. The latter authors believe that the structure is less well organized than in the complete virus and may resemble a pile of discs rather than a flat helix.

The other contribution of interest in this field is that of Markham and Smith (1949) and Markham (1953) on the constitution of turnip yellows virus. This virus, which is produced in very high yield in infected plants, gave preparations that could be crystallized and which were homogeneous in the electrophoreter.

When, however, these preparations were tested in the analytical ultracentrifuge, two boundaries developed indicating the presence of 'top' and 'bottom' components of differing size or density. Methods of separation by ultracentrifugation were devised and it was found that the top component differed from the bottom one in being composed of protein without nucleic acid, in having no infectivity and being a much poorer antigen in rabbits. It resembled the complete virus in being indistinguishable in electron micrographs, having precisely the same crystal form and forming mixed crystals with complete virus and reacting identically with antisera produced by immunization with the complete virus. Standard preparations from infected sap contained up to 40% of particles lacking RNA. It should be mentioned, too, that this virus in its complete form contains 35% of RNA, the largest content of any virus on record. Other plant viruses contain amounts ranging upwards from the 6% of tobacco mosaic virus.

The most interesting feature of these results is that 35% of nucleic acid can be fitted into the protein unit concerned without more than trivial change in the dimensions of the unit or its surface charge. Bernal and Carlisle (1948) indeed found that X-ray studies indicated in the infectious RNA-containing form a repeat unit of 228 Å. while in the non-infectious form the unit was slightly larger, 238 Å. This provides some rather striking indirect evidence for the contention made earlier that protein may be synthesized around a nucleic acid core under a certain stress and subsequently liberated. If RNA does serve as a template for protein synthesis, it may be important to consider the possibility that the protein may be synthesized around a nucleic acid core. This seems more probable than the current idea of a linear arrangement of template and the developing polypeptide chain.

INFLUENZA VIRUS MULTIPLICATION

For the reasons indicated above, we are precluded from attempting to make use of the very extensive data in regard to bacterial viruses. Amongst the animal viruses there would be much to be said for using one of the small viruses like polio, but unfortunately there is still little on record in regard to the chemical constitution of any of these viruses. Influenza virus is now the most extensively studied of all the animal viruses and is also the one with which we are most familiar in this laboratory. If for the time being we omit any consideration of the long filament types and consider only standard well-adapted virus strains like PR8, MEL and WS, a fairly adequate description of the infective virus particles may be given.

Morphologically they form slightly irregular spheres, the diameter of which in electron micrographs depends on the method used to make the preparations. Where adsorption to red cells from fresh fluid virus is used, there is a considerable range in apparent diameter of the virus particle. It is obvious that they are not turned out to a rigidly organized pattern and they would never pack into a symmetrical crystalline arrangement. In many preparations elongate forms looking like two or three fused particles may be seen, and extensive search will usually produce a long filament or two. The mean diameter of the particles is of the order of 100 mμ.

Chemically the virus particles contain: lipid (cholesterol phospholipid and fat), 20–30%; protein, about 60%; carbohydrate, about 10%; and ribonucleic acid, 0·8%; all in terms of dry weight. The RNA contains the normal purine and pyrimidine bases and does not show the 1:1 ratios for adenine:uracil and guanine:cytosine that were found for yeast nucleic acid by Elson and Chargaff (1954). The lipids have the same general composition as host cell lipids and the polysaccharide contains mannose, galactose, fucose and hexosamine and is, therefore, probably of host origin. Knight (1946) has shown that the virus particles contain antigen specific for the host in which the virus has been propagated and presumably of host protein origin. These analytical findings are probably indistinguishable from what would be obtained in similar studies of host cell cytoplasm. They provide at least superficial justification for the claim that influenza virus is 'no more than a fragment of diseased cytoplasm'. Yet it is obvious that the infective particle of influenza virus is something much more definite than that. It has a specific serological character which may well be composed of a number of antigenic components. It has an equally well-defined enzyme (a mucinase) with a consistent relationship to other surface components which allows the differentiation of strains in relatively specific fashion by a comparative study of their hæmagglutinating activity and how it can be inhibited. Finally, it can be shown that influenza viruses have genetic systems which can interact with one another so that mixed infections may produce virus with qualities derived from both 'parents'.

Clearly, even if influenza virus is chemically a mere fragment of host cytoplasm, it carries within it a large number of highly patterned macromolecules which determine the behaviour of the virus and whose replication in the host cell provides another major problem and opportunity in the field of protein biosynthesis.

We shall assume for the sake of discussion that influenza virus particles possess a surface which, in part at least, is composed of a mosaic of protein molecules which carry serological, enzymic and adsorptive determinant groups. It is immaterial to the argument whether each molecule carries all the specific determinants or whether they are distributed in one way or another over a variety of different molecules.

Strictly speaking, there is no evidence for or against the hypothesis that these surface molecules represent the soma of the virus and that, as in all higher organisms, there is a central genetic control probably mediated by nucleoprotein. In all recent discussions on the genetic behaviour of influenza virus (Burnet, 1954*a*) we have adopted this view of a soma and a genome, mainly because it is the conventional and perhaps necessary way in which to express the genetic phenomena in fairly compact and convenient form. The only reasonably direct evidence yet available for its correctness is Ada and Perry's (1955*a*) recent finding that the proportion of RNA in purified virus varies directly with the degree of 'completeness' of the virus sample. This naturally suggests that there is a ⌜nucleic acid, protein⌝ mechanism which is necessary for the normal functioning of the virus as replicating agent. It has still to be proved conclusively that the virus RNA is specific for the virus and not a mere contaminant. The most direct approach would be to show that the character of the RNA differs consistently from one type of virus to another. Ada and Perry's preliminary results (1955*b*) indicate that such differences can be found in the ratio of adenine:uracil which differs significantly between A and B strains of influenza virus.

The close association of RNA with all other types of protein synthesis, and its integral association with all the macromolecular plant viruses makes it highly desirable to direct the initial stages of the discussion to the significance of RNA in the multiplication of influenza virus. Parenthetically it may be added that the results of chemical studies of polio virus which should shortly become available will be highly significant in this context. The only small virus yet tested (Eastern equine encephalitis) contained RNA with no DNA. Polio viruses, however, are considerably smaller, probably have no lipid component (judged by their resistance to ether) and pack into symmetrical aggregates. If they are found, as I think all virologists expect, to be composed of protein and RNA, then the present discussion may well stand. If, however, it is found and confirmed that there is no nucleic acid or DNA only in polio virus, much of the discussion will undoubtedly have to be recast.

1. *Nucleic acid in relation to influenza virus*

The first studies of purified influenza virus led to the conclusion that both RNA and DNA were present although in relatively small amounts. The published figures, however, varied widely amongst themselves, and Ada and Perry (1954*b*), have given what we believe to be adequate reasons for discounting the presence of DNA and giving a lower figure, just below 1% of dried weight, for the RNA content.

The next point of interest is the nucleic acid content of the soluble *Complement-fixing Antigen* (CFA). This is present in large amount in tissues in which the virus is multiplying and, with some reservations, it can also be extracted from sonically disrupted virus particles. CFA can be identified by its serological reactivity which is species specific rather than type specific, and its much smaller particle size than the virus itself. It is normally prepared by absorbing out the virus particles from an emulsion of infected tissue by treatment with red cells in the cold.

Both Hoyle *et al.* (1954) and Ada and Perry (1954*a*) agree that the purified material contains nucleic acid. Ada's findings, however, were at first difficult to interpret, as the proportion of RNA and DNA varied according to circumstances. In material harvested at the earliest practicable period after inoculation, RNA predominated, but at the standard time of harvesting the infected chick embryo lungs, DNA was present in relatively large amount. At this period there was gross cellular disintegration and a large migration of leucocytes to the lung. Ada and Perry's (1954*a*) conclusion was that the CFA probably contained RNA as an intrinsic component and that the DNA found in standard preparations was of extraneous origin. It should be pointed out, however, that the DNA-containing material appeared to be specifically precipitable by the corresponding antibody in rabbit serum. The possibility that under different circumstances both types of nucleic acid have a 'right' to be present in CFA may need to be held in mind.

Hoyle (1953) has suggested that CFA is simply the genetic material of the virus, that in effect the virus particle is a group of CFA units enclosed in a bag of lipid derived from the host cell surface. Without going quite so far, it still seems necessary to look on soluble CFA as an index of the presence of the multiplying pool of virus components in the infected cell. It may well be that soluble CFA is heterogeneous, as was in fact indicated by Ada *et al.*'s (1953) studies using agar gel precipitation reactions. Some suggestions explaining how this heterogeneity might arise are made later, but in any case if there are self-replicating protein-RNA complexes in the infected cell they would be expected to make an important contribution to the antigenic qualities of extracts of infected tissue. It may be justifiable then to make a tentative identification of at least a portion of the CFA units with specific protein-RNA complexes which may well take an essential part in the process by which virus components are replicated during the intracellular phase.

Finally, we should mention the histological appearance of cells of the allantoic cavity infected with influenza virus and stained with pyronin and methyl green (Unna-Pappenheim). Bates's unpublished studies in my laboratory showed a well-marked sequence of events. If one adopts the normal convention of ascribing pyronin staining to RNA and staining with methyl green to DNA, the salient features are as follows. In the early stages of infection there is a diffuse accumulation of RNA in the cytoplasm which becomes more granular in distribution and accumulates particularly in the perinuclear region. In the later stages one finds many cells with a dense red-staining material surrounding the nucleus. Only a proportion of cells show nuclear changes beyond enlargement of the nucleolus, but these changes are striking. Typically one sees complete disappearance of the nuclear membrane with a number of round or irregular masses of green-staining material to represent the DNA of the nucleus. There were some hints that this might be the result of infection in a cell undergoing mitosis, but this can be no more than a speculation at present.

While considering the significance of the nucleus in infection with influenza virus, it is relevant to mention the consistent finding of Watson and Coons (1954) that in the early stages of infection in the chick embryo there was evidence of influenza virus antigens in the nucleus. This was absent in the later stages when, by the fluorescent antibody technique, the virus was present only in the cytoplasm.

Probably the strongest evidence against DNA taking any active part in the process of virus multiplication is to be found in Stevens's (1954) work. He showed that neither nitrogen mustard nor heavy doses of X-rays inhibited the production of virus in de-embryonated eggs.

On the whole it seems clear that RNA must play a highly significant part in influenza virus multiplication and, with some reservation, very unlikely that DNA is directly involved.

2. *An attempted visualization of the structure of influenza virus particles*

Before attempting any further discussion of the process of influenza virus multiplication it is probably desirable to try to summarize the data so far discussed by an attempt to visualize the structure of the infective particle. Such a picture must be based (i) on the established physical and chemical qualities of the particles, (ii) on the evidence that many of the components are unmodified host cell components, (iii) on the presence of numbers of highly specific macromolecules in the surface of the particle, (iv) on the existence of recombination with its implications for the nature of genetic determinants, (v) on the presence and specific character of RNA in the virus, and (vi) on the occurrence and character of soluble CFA.

In the first place an estimate may be attempted of the approximate number of specific macromolecules concerned.

An influenza virus particle has a 'molecular weight' of approximately 500 million. On Ada's results it might contain 20 RNA macromolecules of molecular weight 250,000 each, and if two-thirds of its surface were composed of protein molecules around 100,000 molecular weight there would be about 300 of these. This would only be about one-tenth of the protein in the particle which probably includes a thousand molecules or so of host protein. Lipid molecules packed vertically, perhaps in a double row, might fill the interstices between the specific protein molecules of the surface. The possibility of some more intimate association of protein and lipid as lipoprotein may also need consideration.

A RNA complex of molecular weight about 250,000 associated with a few molecules of protein would give a particle of the order of size of the soluble CFA. We assume that there is something of the order of 20 such complexes in the infective particle and, further, that these complexes are the essential replicating mechanisms and that, in a rather definite sense, they are the real virus. It must be left open whether these complexes are identical with, or have any significant relationship to, the soluble CFA that is extractable from infected cells.

In line with the picture of protein synthesis developed from other fields, one pictures a rather direct relationship between the RNA and the functional protein it produces or, in other words, between the genome and the soma.

To take the simplest possible hypothesis, we may assume that each surface protein molecule carries enzymic, adsorptive and serological determinants and that the nature of these is determined by the template carried by a protein-RNA + pair. On this view, the specific molecules which form the virus particle by arranging themselves in the surface are liberated 'partial replicas' of one of these protein-RNA + complexes. A development beyond this simplest structure would be to have, say, two or three types of RNA + units per virus particle, each responsible for a different specific type of protein molecule. A stable structure would presumably only be possible if the two or three types of surface molecule arranged themselves with host components in an appropriate mutual arrangement.

This picture of the infective particle of influenza virus is represented diagrammatically in Fig. 6. It has, in fact, been built up very largely from considerations that have not yet been introduced into this discussion but, in an attempt to increase the clarity of the presentation, it seems better first to develop a suitable model of the virus particle and then to see how the concept that it represents fits the wide range of

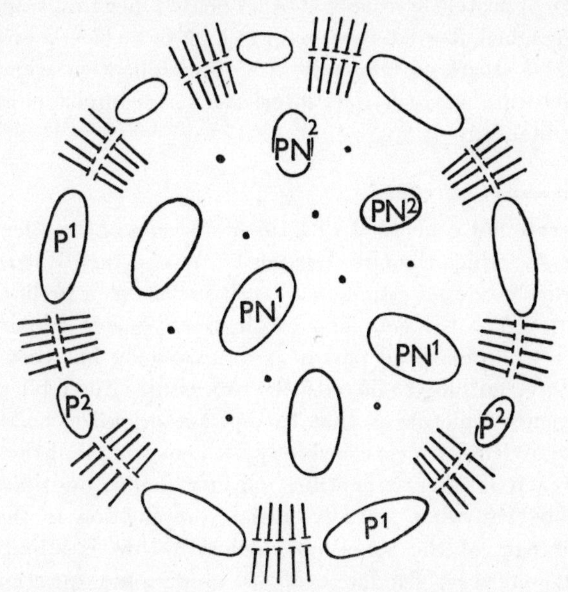

FIG. 6. A highly schematized diagram of the structure of an influenza virus particle. The surface is composed of specific virus protein and of lipid derived directly from the host cell. Within are the nucleic acid (RNA) and protein complexes which make up the virus genome, together with casually accumulated material which may carry patterns derived either from host or from virus. P, protein; PN, ⌜protein, nucleic acid⌝ complex.

experimental facts that have been recorded about influenza virus multiplication.

The first requirement is to extend the picture to cover the processes that must take place in the multiplying phase of the virus in the infected cell. This will necessitate an attempt to grasp the implications of the eclipse phase and the other quantitative aspects of virus multiplication and then to look for evidence that will serve as a guide to the formulation of hypotheses of replication that will be in accord with what is known of protein synthesis in other fields. Once that stage has been reached, it will be necessary to look more closely at three special features of influenza virus multiplication which we can indicate succinctly as interference, incompleteness and recombination.

3. *Process of infection*

We are not concerned with the first stages of the infectious process in which the virus particle becomes attached to muco-polysaccharide receptors on the cell surface as a preliminary to entry into the cell. The surface components responsible represent an essential part of the virus soma, but they may well have nothing to do with the process of virus replication. Our general picture is that the surface organization of the virus particle is concerned simply with ensuring that the virus can reach another susceptible cell in which to multiply. We believe that what actually initiates replication in the cell is *not* part of the soma—a conclusion that is admittedly based on very slender evidence and some questionable analogies.

It appears to be an adequately established fact that some time after entry into the cell, influenza virus becomes undetectable as infective virus. All concerned with electron microscope studies are agreed too that no morphological evidence of virus particles is visible in the cytoplasm of infected cells. New particles only become evident as they emerge from the surface of the cell. The virus appears to disintegrate into smaller units, perhaps under the influence of intracellular enzymes, perhaps for some other reason. Three or four hours later new virus begins to be liberated so that a process by which everything needed for the fabrication of new particles is produced must be set in motion without delay. The facts of recombination make it clear that new virus particles are drawn in some sense from a pool of components derived from the one or more virus particles responsible for initiating infection in the cell concerned.

In terms of the picture we are developing, the central features of influenza virus multiplication are (i) the replication, directly or indirectly, of virus components brought in by the initiating infective particle to give rise to what we have called the multiplying pool; and (ii) the fabrication from pool components of the infective particles of the new generation of virus.

The evidence available in regard to the nature of the multiplying pool is very limited. In fact, all that can be said is that in extracts of infected cells early in the process there is evidence to suggest that soluble CFA is present before hæmagglutinating or infective particles and that relatively early extracts show a higher proportion of hæmagglutinin to infective virus than is present in virus liberated in normal fashion. It is unfortunate that so far it has not been possible to make effective use of differential labelling of somatic and genetic components of the virus analogous to the methods used by Hershey and Chase (1952) for their studies of the entry and multiplication of bacterial viruses. Under the circumstances any interpretation of what is happening in the pool must be tentative and drawn chiefly from analogies with other systems in which protein and nucleic acid are being synthesized.

We shall adopt, therefore, the same basic hypotheses as have been used in discussing the synthesis of adaptive enzymes or antibodies, and speak of RNA+ units as providing templates for the synthesis of specific protein and again under appropriate conditions accepting protein as the template for new specific RNA+ synthesis. If we accept Ada's evidence for specific differences in the relative amounts of the four bases in RNA from A and B strains respectively, then we must postulate that RNA brought in by the initiating particle plays a highly significant part in the replication of virus pattern. We believe, therefore, that the simplest way to visualize what is happening is to picture the pool as containing protein-RNA + complexes which, according to circumstances, are capable of producing replicas of either protein or nucleic acid. In all probability, the stimulus in either case is the presence of adequate amounts of the amino acids or nucleotides (or smaller units) from which the macromolecules can be synthesized. There are hints from other directions that the mobilization of ribonuclease to break down inert RNA may be a specially effective stimulus for the production of a new supply of building blocks and of new RNA.

We can be certain that virus functional protein is being produced in increasing quantity, that RNA is increasing in the cytoplasm of the infected cell, and, if we accept Ada's figures as representing functional virus RNA, that this too is being synthesized in increasing amount. As hinted earlier, we are inclined to think of the RNA + template as something more complex than a linearly arranged code against which a 'corresponding' sequence of amino acids is laid down. The linear 'alphabet' of the polynucleotide is so much simpler than that of the polypeptide that such an arrangement seems to be impossible. But if more than one polynucleotide chain can play a part in determining the sequence of amino acids, a much more complex code becomes possible. In a very crude fashion Fig. 7 illustrates one possibility by which a polypeptide could be synthesized according to a code depending on the linear arrangement of bases in a polynucleotide and on the way the foldings of the chain bring different groups into relationship. There is no evidence whatever that this is the actual way in which the two are related, but it illustrates the sort of way in which, for instance, the nucleic acid might be related to the specific protein of a virus like turnip yellows. The requirements are that the protein should be superficial, since it is the serologically reactive component, and that the two components are separable with relative ease.

With a more complex relationship like that in Fig. 7 at the back of our mind it is, however, still legitimate to picture the situation more diagrammatically as linear units in apposition, whose function is rather to produce protein (P) replicas than nucleic acid (RNA+) replicas. The P macromolecules are then available for the construction of the virus surface (see Fig. 6).

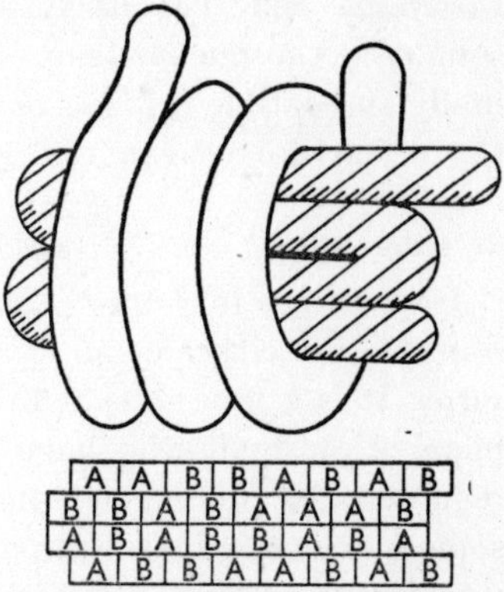

A	A	B	B	A	B	A	B
B	B	A	B	A	A	A	B
A	B	A	B	B	A	B	A
A	B	B	A	A	B	A	B

FIG. 7. A diagram to suggest the increased versatility of a binary code when used in the fashion described in the text. If the relationships of A's and B's to immediately adjoining symbols are included, many more potentially meaningful arrangements are available than if the binary symbols were in a simple linear order.

Again with the proviso that we are merely trying to develop the simplest possible picture of the replication of influenza virus, we can use the simple protein-RNA+ system to build a model from which working hypotheses may be derived which will suggest a new experimental approach.

When the conventionalized particle that we described in the last section enters the cell, we can regard it as dissolving in the cytoplasm, thereby bringing the RNA+ units directly into the intracellular environment. If the cell is susceptible, these units will find it possible to build up a pool of virus components from which eventually new virus will be produced. But it is of the essence of the argument that the cell is a susceptible host cell. It is not sufficient that the virus should dissolve in the cytoplasm of an actively functioning cell. There must be some special attribute that makes it a *susceptible* cell. The likely clue to this attribute must be sought in the phenomenon of interference by which a normally susceptible cell can be rendered insusceptible through the action of inactivated virus of similar type.

4. *Interference*

Interference is characteristically induced by the prior administration either of an avirulent living strain or of killed virus. It is a process that takes some hours to become complete. Almost all who have theorized on the subject have regarded the process as essentially due to the blocking of some key intracellular component by virus components, presumably the protein-RNA+ units, which lack the power to compel the supply of host nutrients. One must postulate a specific complementary attachment of some sort; presumably a pattern on the surface of the virus unit finds a limited number of complementary patterns available in the accessible regions of the cell. It is conceivable, but in the case of influenza virus rather unlikely, that the virus unit must make the necessary association with a complementary cell unit in the nucleus, the replicating mechanism then passing to the cytoplasm. This appears to be the case with herpes virus infections, but here the presence of DNA in the virus differentiates the situation sharply from influenza virus replication. Like so much speculation at this level, the requirements have to be left broad and expressed as interaction of patterns without specific indication as to where the reaction occurs. There are two requirements in any interpretation of the nature of interference.

(1) Union of the two complementary patterns must provide a necessary (but not sufficient) condition for the provision by the cell of means which will allow replication to take place.

(2) Since inactivation by ultra-violet light is a very convenient way of producing an interfering agent and since the effective wave-lengths are those of nucleic acid absorption lines, it is probably necessary that the nucleic acid brought in by the virus be intact if multiplication is to become possible; this also implies that the protein element is the carrier of the pattern needed for contact with the cell unit.

These requirements can be satisfied only by postulating the existence of preformed host cell units or organelles carrying a pattern complementary to part of the virus replicating unit. Such a Cell Pattern can be referred to as a CP and we have suggested that it is a portion of the specific virus protein that is responsible for the specific union to the CP. Only when this union can take place is it possible for virus replication to proceed, but, in addition, some other quality or component must be brought into the cell by the infecting particle if the next stage of the process is to go on. This second quality is easily destroyed by ultra-violet light or moderate heat, without loss of the first. Up to the present no serious suggestion has been possible to explain what constitutes this second quality of viability or full infectivity. There are, however, a few hints on the nature of the process to be obtained by considering some further experimental results in the light of the concepts so far developed. Except where otherwise indicated, all the examples are concerned with interference induced by virus 'gently' inactivated by heat or ultra-violet light, of the same type as the challenging influenza virus.

(1) For the full development of interference a period of time of the order of the time of first liberation of new virus is needed.

(2) Full interference appears to be inducible in an otherwise susceptible cell by the entry of a single effective non-viable virus particle (Fazekas and Edney, 1952). Time would clearly be necessary for the entry of the interfering particle, its disaggregation and the movement of the various components in the cell until they find and block a majority of the cell units that can function as 'feed-in' units for *active* virus components. Some doubt must remain as to whether interference can be induced directly by what enters with the virus or whether limited replication of at least some of the components is needed. Analysis of the amount of CFA produced by various non-viable inocula might provide an answer.

(3) Influenza virus in certain circumstances can produce interference against quite unrelated virus, e.g. in the mouse brain against Western equine encephalitis virus (Vilches and Hirst, 1947), and on the chorion against vaccinia (Depoux and Isaacs, 1954). Nothing much can be said in view of the fact that even less is known about interference with other viruses than about that with influenza. Perhaps the more important question is why influenza does *not* provoke interference against many other viruses. The general lines of the answer will probably be that virus units in our present sense will eventually be classified in terms of the host unit to which they make specific adsorption.

(4) Henle *et al.* (1947) showed that when an allantoic infection had been induced with an A strain, and a large dose of inactive B virus was added in an hour's time, there was no interference with the process of liberation, but virus production was limited to the first cycle. This is only what would be expected. If, however, a large dose of *homologous* inactivated virus was given, there was gross interference with the first cycle yield of virus.

This requires a good deal of elaboration of our hypotheses. In the first place it calls for the introduction of the idea that, as the multiplying pool increases, there will have to be a concomitant increase in the number of CP's available. The double process of contact with CP plus the influence exerted by intact virus RNA must be presumed to set the local situation for replication of both virus and host components, involving no doubt a mobilization of amino acid and nucleotide pools. From Henle's results it appears that, in a proportion of instances, the subsequent entry of a non-viable virus particle of the same type will prevent the continuation of a replication process already initiated. The activity of homologous but not heterologous virus must surely signify that CP units newly produced under the stimulus of an active virus unit are specifically related to the homologous virus component and less so to the heterologous one. This might mean that the first new units of pattern produced by the cell have a complementary pattern actively impressed on them by the virus component. Some pre-existent pattern more or less complementary to a variety of heterologous virus components must be postulated ; the suggestion from Henle's experiment is merely that the first replications of this pattern under virus influence give patterns that seem to be more specific for the homologous type of virus unit. It seems hardly possible to be sure whether the increased specificity is a selective effect or an indication of active modification.

5. *Incompleteness*

When a classic influenza strain like PR8 is propagated in the allantoic cavity under standard conditions from a very small inoculum, the infective fluid obtained shows a characteristic ratio between its infective titre ID_{50} and its hæmagglutinin titre. With the conventional techniques used in Melbourne the ratio is normally about $10^{5\cdot6}$, i.e. a fluid with a hæmagglutinin titre of 10^{-2} will have an ID_{50} of $10^{-7\cdot6}$. If, however, following von Magnus (1952) we carry out daily passages by transferring 1 ml. of undiluted fluid, it will probably be found that the third passage fluid has an ID_{50} to hæmagglutinin ratio of only about $10^{2\cdot6}$. At first glance it appears that this fluid contains only one infective particle amongst 1,000 which are capable of hæmagglutination but not infective. Such a fluid is spoken of as 'incomplete virus', the degree of incompleteness being specified by the deviation of the ID_{50} to hæmagglutinin ratio from the standard value.

Such functionally incomplete virus can hardly be distinguished from a virus treated with moderate doses of ultraviolet light or with exposure to minor degrees of heating. There is, however, a consensus of opinion that the particles are physically and chemically different from standard infective particles. There are some discrepancies amongst the reports, but it has been claimed by one author or another that an incomplete virus shows (i) larger, flatter and less uniform particles in electron micrographs (Werner and Schlesinger, 1954), (ii) lower sedimentation rate in the ultracentrifuge (Gard and von Magnus, 1946), (iii) a higher content of lipid (Uhler and Gard, 1954), (iv) a lower content of RNA (Ada and Perry, 1955*a*). The ratio of particles (counted by electron microscopy) to hæmagglutinin titre is not significantly different from that of normal virus.

A third method by which the activity of a virus fluid can be measured is to determine the amount of hæmagglutinin produced in the first cycle of growth by serial dilutions. The de-embryonated egg is convenient for the purpose and, using the technique described by Burnet, Lind and Stevens (1955), the yield of hæmagglutinin at 7 hr. can be plotted against the inoculum also expressed in terms of hæmagglutinin. In this way one can obtain an index of the relative power of two fluids to produce hæmagglutinin under these conditions. If we take a standard virus and a highly incomplete virus fluid and test them for hæmagglutinin titre, for infective titre by limit dilution in the allantoic cavity, i.e. ID_{50}, and for capacity to produce hæmagglutinin in first-cycle infections, we may find that the incomplete virus has the same hæmagglutinin titre, produces about one-tenth of the hæmagglutinin in first-cycle yield and has only $\frac{1}{3000}$ of the infectivity by allantoic titration. It seems superficially that, of 3000 particles, all can adsorb to a red cell surface and hæmagglutinate, 300 can enter a susceptible cell, initiate multiplication and liberate a full quota of hæmagglutinin which, however, has no further capacity to infect, and one only can initiate a series of continuing infections through successive cell generations.

These are the essential features of incompleteness as seen with strains PR8 or WS. It is clearly a phenomenon that must in some way be fitted into any picture of influenza virus replication.

The characteristic feature of incompleteness is its vicious circle character. Virus grown under circumstances involving multiple primary infection of cells is more incomplete than normal; when this virus in its turn is passaged in the same fashion, it becomes still more incomplete and so on. Virus that is capable of one cell-generation of replication fails to survive a second.

Some quality *X* seems to be concerned which, if present in adequate amount, allows the full replication of all virus components, including itself. When it falls below a certain threshold, however, replication, first of *X* and then of other components, becomes progressively less effective. It seems, too, that *X* is one of the most readily damaged components of the virus.

This *X* unit must clearly be closely related to or the same as the quality of viability or infectivity that we had to postulate in discussing interference. Only in its presence does contact of the virus unit to the CP's of the host result in the activation—perhaps the mobilization of 'building stones' for both protein and nucleic acid—that is needed for adequate replication of virus. *X* then seems likely to be some quality or portion of the virus RNA and this, of course, is in harmony with Ada's finding that incomplete virus is deficient in RNA. The simplest possible interpretation of incomplete virus is that it is the result of a process by which the ratio of specific protein to RNA progressively increases. If such a process is possible, its analysis should throw light on the nature of the interaction of RNA and protein in synthetic processes generally. Something broadly analogous to incompleteness may be seen in Cohen and Barner's (1954) studies on a thymine-requiring mutant of *Escherichia coli*. Such a strain is capable of cytoplasmic growth and even of the development of adaptive enzyme in the absence of thymine. In the absence of DNA synthesis this unbalanced growth is associated with irreversible loss of the power to multiply. The parallel is by no means close, but indicates the possibility of unbalance being of major significance.

At this point then we may return to a consideration of what is probably happening in the multiplying pool. We have assumed that there are present in the invaded cell a certain number of cell patterns (CP) with which virus components, probably protein, can react and, depending on circumstances, influence them in one of two ways, (*a*) by activating the CP to form with the virus unit a centre of synthetic activity leading to the replication of both host and virus components, (*b*) by blocking the CP and denying it to virus components otherwise capable of inducing (*a*). Processes of this type must continue as long as the multiplying pool persists. There is a continuing production of virus protein molecules, of virus RNA, of host RNA and associated structures and a mobilization of the necessary 'building stones'. It is probably immaterial whether the CP's are expressed in protein or RNA, but it is simplest to keep the situation symmetrical and assume that the CP is expressed in RNA.

In the multiplying pool we can imagine that, as each new macromolecule of virus protein is produced and liberated into the pool, it has three alternative immediate fates: (i) it may unite with a host CP, so denying this to subsequent activation; (ii) it may serve as a template for the reconstitution of a RNA+ unit; or (iii) it may remain in the pool to be available for subsequent fabrication of infective virus particles. The new protein-RNA+ unit produced in (ii) will have two possible fates: (*a*) it may make contact with a CP and initiate a new series of replications; or (*b*) it may also wait in the pool for eventual incorporation into an infective particle. In this way a rather delicate balance between interference and free multiplication of virus components will allow relatively minor circumstances to make a great difference to the functional composition of the pool.

The requirements of the results of recombination experiments to be discussed later demand that there should be at least two types of virus genetic unit—'linkage groups'—and we may picture the multiplying pool as containing, in addition to host cell components and products, two sorts of specific virus material. The protein molecules we can call P^1 and P^2 and the corresponding protein-RNA complexes, P-RNA^1 and P-RNA^2. The possibility of a third set of genetic qualities cannot be wholly excluded.

The most reasonable picture of the formation of the infective particles from the pool is that it results from the interaction of a high concentration of virus P molecules and the cell's free surface. Here, perhaps assisted by the specific mucinase of the protein molecules, the surface of the virus to be takes shape as a mosaic of specific virus protein and cell surface lipids and mucoproteins. Depending on circumstances, the surface defines the morphology as a filament, a string of beads or spherical particles. In the process a fragment of cytoplasm containing host and virus pool components is enclosed within the active surface. Liberation of virus on this view will depend on the development of an adequate concentration of the intracytoplasmic virus pool at the points where it abuts on the free surface of the cell. If, as a result of the process of blocking, an adequate concentration cannot be built up, no virus will be liberated. The content of genetic material in each particle will be determined simply by its concentration in the pool at the surface where the virus has been produced. In the process of defining the morphology of the particle by surface active molecules, a portion of the cytoplasmic pool will, as it were, be trapped within the particle. Its content will determine the number of genetic units that are incorporated.

On this view, incomplete virus is composed of those particles which, in their formation, trap an inadequate supply of specific RNA. To make the position clear, it may be advisable to adopt a set of hypothetical numerical values for the ratio of protein to protein-RNA complexes, with the proviso that functionally speaking a P-RNA complex becomes only equivalent to P if the RNA has been damaged by ultra-violet irradiation or by simple thermal degradation.

Suppose then the normal ratio of P:P-RNA is 50:1 and that when the ratio rises to 250:1 the blocking that follows entry into a susceptible cell is too much to allow any subsequent liberation of virus.

The usual method of inducing gross incompleteness is to infect cells with an average of two or more virus particles and we may picture what happens in such a case as follows.

When a normal particle and one which by thermal degradation has a ratio of 150:1, say, together enter a cell and replication commences, there will be sufficient blocking to produce a serious deficiency of P-RNA in the pool. Taking an average of the yield from very many such cells we may find that only 0·1% of the particles liberated have a ratio below 100, 10% have ratios 100–250 and 90% above 250.

If a single particle with a ratio of 100 or less enters a susceptible cell, it will at least maintain this mean value in the pool and amongst the next generation there will be many particles with considerably lower ratios. These will have a survival advantage in subsequent generations so that if an embryo is infected with such a particle the end result will be a full titre fluid with particles predominantly of the normal 50:1 character.

When 150:1 particles infect, however, there is sufficient stimulus for replication to take place, but the ratio progressively worsens and when the virus is liberated the mean ratio will be perhaps 400:1. This virus will be measurable as hæmagglutinin, but on entry into other cells the virus will fail to replicate. Particles with this range will, therefore, represent the single-generation type of virus not detectable when inoculated as a single infective dose, but contributing hæmagglutinin in a single cycle experiment.

Any particles with a P:P-RNA ratio above the critical level which we have arbitrarily fixed at 250 will give no evidence of their capacity to multiply under any circumstances.

Strains of influenza virus differ greatly in the ease with which they produce incomplete virus, the old strains PR8 and WS seeming to be unduly capable in this regard. On the present hypothesis this might be explicable in two ways: (i) that only in the PR8 type of strains is it possible for protein production to be so completely divorced from P-RNA replication; or (ii) that with long adaptation of the virus to the chick embryo, host RNA can serve more directly for virus protein replication, but is not of such a nature that it can itself be replicated under the conditions of infection.

It may be convenient at this stage to note the experimental findings in regard to incompleteness in summary form and to add some notes as to the relevance of the current hypothesis to each point which has not already been adequately covered.

(1) The readiness with which incompleteness can be induced varies from one strain to another.

(2) Incompleteness is essentially an artificial phenomenon encountered only in first-cycle multiplications. When a virus multiplies to full titre from a small inoculum the final yield must have come largely from doubly infected cells yet it is virtually complete. This depends primarily on the fact that, where opportunity exists for two or more cycles, fresh, fully active, complete virus particles have a great survival advantage over any damaged forms. It is probable that the process of harvesting and chilling diminishes the 'vigour' of a particle so that it experiences more difficulty in initiating infection than a newly liberated one. A large proportion of even the best inoculum will, therefore, be at a disadvantage of some sort. This may well fit into Fazekas and Graham's (1954) finding that the degree to which incompleteness can be developed is directly related to the time required to initiate first-cycle infection. Throughout this period thermal degradation of the virus particle is going on.

(3) Functionally, an incomplete virus cannot be easily differentiated from virus that has been partially inactivated by heat or ultra-violet light. Von Magnus, however, finds that it is not possible to reproduce with thermally inactivated virus the characteristic growth curve in the chick embryo.

There can be no question that incomplete virus is different from complete virus inactivated by heat. Even Horsfall admits this, but it is probable that a virus particle with damaged RNA behaves very like one lacking RNA.

(4) Incomplete virus is deficient in RNA (Ada and Perry, 1955*a*) and, according to Uhler and Gard (1954), contains an excess of lipid. Ada (personal communication), however, finds the latter is not reproducible.

We should regard the lipid content as depending essentially on adventitious circumstances, including perhaps the degree of disorganization of cytoplasm and cytoplasmic surface at the time the virus is liberated.

(5) There are degrees of incompleteness most clearly shown by the discrepancy between estimates of infectivity based on LD titrations and on the production of first-cycle hæmagglutinin.

(6) Incomplete virus readily contributes genetic characters to recombinants with active virus (Burnet, Lind and Stevens, 1955). This is analogous to the similar capacity found with heat-inactivated virus (Burnet and Lind, 1954*a*).

This has already been discussed, but its importance should be stressed again. It is hard to escape the impression that a pattern carried only on protein can, under suitable conditions, be re-established in P-RNA complexes. The alternative is one that has emerged occasionally already, viz. that all the *pattern* is carried in the protein molecules and that the virus RNA acts only as a primer of autocatalytic character. The facts of interference, however, point rather strongly against it and even more so the specific differences in RNA from different species of influenza virus.

(7) Virus with the character of incomplete virus is obtained in the early stages of virus multiplication when extracted from the infected tissue.

We should lay little stress on this simply on the basis that very newly formed virus may not have 'settled down' into its appropriate pattern and may, therefore, be subject to damage in the process of extraction. It cannot be taken as evidence that the RNA is the last thing to be inserted.

(8) Incomplete virus in a susceptible tissue (Beale, 1954), like complete virus on the chorion (Isaacs and Fulton, 1953), can give rise to CFA without liberation of virus.

CFA contains specific protein of a pattern foreign to the host and RNA which according to Ada (private communication) shows a pattern of bases very close to that of RNA from the host tissues and unlike that of the RNA found in purified virus particles. A probable interpretation is that it represents some sort of primary mobilization of host synthetic activities from which is derived material used for the replication of definitive virus components. It still remains open to suppose, however, that CFA may have more than one origin. Protein with virus-determined pattern combined with RNA could be (i) the hypothetical P-RNA units of the infective particle, (ii) virus protein associated with host RNA units, e.g. microsomes, (iii) virus protein fortuitously adsorbed to nucleic acid liberated by mechanical disintegration of the cells during extraction. If we exclude (i) on Ada's findings, any of the other hypotheses would allow CFA production with minimal virus liberation.

6. *The dynamics of influenza virus multiplication in the allantoic cavity*

One of the classical problems of theoretical virology is the interpretation of the process of influenza virus multiplication in the allantoic cavity of the developing chick embryo. A great amount of quantitative data is available, but within the last year papers by Horsfall (1954) and by Donald and Isaacs (1954) have necessitated a considerable reassessment of these earlier results and there are still some outstanding discrepancies. For the time being it is probably advisable not to regard current estimates of the ratios, for instance, of (infective dose) : (hæmagglutinating unit) : (visible virus particles) as more than preliminary, and to speak in terms of comparative magnitudes only. On general principles one would predict that it will be found eventually that, under all circumstances, the count of morphological particles will be higher than the number of units recognizable by functional methods, whether this is by the capacity of a virus particle to hold two red cells as a dimer, or by the power to induce continuing infection in the allantoic cavity. One might guess that it takes an average of two morphological particles to hold two cells together and about five to provide a 50% chance of inducing infection when standard 40-hr. virus is used. It is probable, as Horsfall contends, that the freshest possible virus just liberated from the cell will induce 50% of infection by a number of particles, on the average, smaller than five; and there is no doubt that, under many circumstances, the relative efficiency of the particles as infective agents is greatly reduced. At least with the strain PR8 there is evidence that much of its efficiency as an infective agent is lost if delay and chilling intervenes between liberation and opportunity for fresh infection. This appears to be the reason why PR8 has a first-cycle time of 8½ hr., while subsequent cycles average 5½ hr. (Fazekas and Cairns, 1952). Instead of entering rapidly into the cell, the virus is held on the surface for 3 hr., on the average, before effectively initiating infection. During this period it may well be subject to further thermal damage and, although still capable of initiating infection, will give rise to a high proportion of incomplete virus. Freshly liberated virus which can rapidly find other susceptible cells to enter will not show this delay, so that the second cycle of cell infection will show a normal period (Fazekas and Graham, 1954). In second and subsequent sets of infection only complete virus will be involved; it will obviously have a high survival advantage over incomplete or otherwise abnormal virus. This is equivalent to stating that incompleteness is wholly a first-cycle phenomenon.

On this interpretation a cell invaded by two or more newly liberated virus particles behaves exactly as if it were invaded by one, except perhaps for reaching the stage of liberation of new virus a little earlier.

The sequence when a small dose of virus is inoculated into the allantoic cavity will be (*a*) adsorption of virus particles to cell surface, (*b*) entry of a proportion of these after a variable time with initiation of infection, (*c*) commencement of liberation of a new generation of virus particles 3 hr. after effective entry. This will comprise a mixture of complete and incomplete virus, but only the complete virus will succeed in producing a line of descent. In its turn this will pass to uninfected cells, to repeat steps (*a*), (*b*) and (*c*), but without the delay in initiating infection characteristic of the first cycle. In the meantime, liberation of virus from the first crop of infected cells goes on for many hours. After a period, the stage is reached when there are more virus particles in the fluid than hitherto uninfected cells. A large proportion of the last moiety of cells to be infected will, therefore, receive more than one invading particle. The general sequence of events will not, however, be modified. The titre of the fluid will reach a stable level when all cells have had an opportunity to liberate a quota of virus particles. Since the hæmagglutinin titre finally obtained is virtually constant irrespective of the size of the infecting dose, each infected cell may be presumed to provide an approximately equal contribution of virus to the fluid. On orthodox calculations (Fazekas and Cairns, 1952), about 100 virus particles are liberated per cell. According to Donald and Isaacs (1954) it may be closer to 1000 per cell.

7. *Recombination phenomena*

By recombination phenomena we refer to the appearance in progeny from a double (or multiple) infection, due to viruses with two or more distinguishing characters, of qualities which must have been derived from both 'parent' strains. In such work we are normally concerned with a series of 'marker' characteristics and the main virtue of influenza viruses is the possibility of demonstrating many differences between markers by simple *in vitro* methods based on hæmagglutination. In the logical handling of the results of recombination experiments we must use genetic conventions and in order to avoid the use of the word 'chromosome' we have adopted the term 'gene thread' to signify the genetic structure that determines the nature of the qualities with which we are concerned. There is really no evidence one way or the other for a linear distribution of genes, and the formulation in the form of such a linear arrangement arises simply from the convenience of writing a series of symbols in that fashion.

The gene thread must have a physical basis in the form of what we have so far referred to as P-RNA complexes or units, and we are inclined to consider each gene thread to be composed of two such complexes and to give rise to two corresponding protein macromolecules. A large-scale analysis of the protein components of virus particles might easily provide an experimental answer to the number of types of protein carrying specific virus characters. It is relatively immaterial to the discussion whether one or more distinct types (apart from allelic modifications) are concerned.

Doubly neutralizable virus, heterozygotes, phenotypic mixtures. The only phenomenon resulting from genetic interaction between influenza viruses that has been fully established in two independent laboratories is the appearance from double infections with serologically distinct influenza A strains of a virus fluid whose hæmagglutinin can be neutralized significantly by antisera corresponding to both the original strains (Fraser, 1953; Hirst and Gotlieb, 1953). Both groups of workers also agree that it is not possible to isolate at limit dilution and maintain strains showing this characteristic. Our interpretation is that the first-cycle population obtainable from mixed infection in the de-embryonated egg is an *unselected* cross-section of the various types of particle that can be fabricated from the replicating pool derived from the double infection. A very large proportion of the particles are not fully viable for reasons discussed above in regard to incompleteness. They are presumed to be purely random fabrications from the building blocks available for both soma and genome. The average particle will have a mosaic of surface units, some of one serological character, some of another: similarly, the genetic units (gene threads) will also be distributed at random and in most cases in too small a number to make the particle fully viable.

For a particle to be regarded as fully viable it is necessary that it should initiate infection when inoculated alone into the allantoic cavity and that its descendants should continue the process of infection until all allantoic cells have been involved and a hæmagglutinating fluid of high titre is available for study. Under most circumstances the characters of the final fluid can be regarded as representing the phenotypic structure of the individual particles making up the final population—and of the particle from which the population was derived. The last, however, will not necessarily hold when we are dealing with an abnormal particle as the primary infecting agent.

Gotlieb and Hirst (1954) have found that fluids showing double neutralization contain a proportion of heterozygotes, in the sense that a high proportion of eggs inoculated with volumes which contain an average of less than one infective dose and giving evidence of infection contain both serological types of virus. Our own work (Burnet and Lind, 1954*b*) has given less direct but equally cogent evidence in the same direction. It is rare, however, to obtain a doubly neutralizable fluid from such limit titrations and we have to believe that, for one reason or another, virus particles of homogeneous genetic character are more viable than gross heterozygotes.

We should guess that this depends on the greater ability of uniform surface units to aggregate into a stable, functionally effective surface, with a consequent differential survival of any such types which emerge from successive replicating pools. It should be stressed that for technical reasons the direct study of heterozygosis and similar phenomena is limited to serological aspects. A serological heterozygote is clearly genetically unstable, but we have no evidence that heterozygosis of other types is similarly unstable. We believe, in fact, that a form of heterozygosis is highly characteristic of the progeny of recombinations between virulent and avirulent strains of the same serological type. The interpretation of the change from O phase to D phase given in a later section assumes that intermediate types which might well be regarded as heterozygotes can exist in a relatively stable form.

Interchange of linkage groups. Over what is now an extensive experience with influenza A virus strains we have observed a phenomenon which we have referred to as interchange of linkage groups. In all genetic work the experimenter is concerned only with *differences* between the types he crosses and within the progeny to which they give rise. For both conscious and unconscious reasons we have tended in our work on recombination to use strains differing in as many conveniently demonstrable qualities as possible. The existence of two highly modified derivatives of the standard strain WS, Stuart-Harris's neuro-WS and our strain WSE adapted to the chorioallantois and highly pathogenic for the chick embryo, provided contrasting types to be 'crossed' with more normal strains which had simply undergone adaptation to growth in the allantoic cavity. With the two strains

MEL and WSE we had available six characters by which they could be differentiated and we adopted the convention of indicating them by the first six letters of the alphabet, upper case letters being used for the qualities shown by MEL, lower case for the contrasting qualities exhibited by WSE. With an occasional exception, the progeny isolated from double infections showed four combinations only of these characters:

ABCDEF, *ABcDeF*, *abCdEf*, *abcdef*,

two of them being the original forms and two being reciprocal recombinants which may be more conveniently expressed as

ABDF-ce (M+) and *abdf-CE* (WS−).

Similarly, neuro-WS is essentially similar to WSE with the additional positive quality of producing encephalitis on injection in the mouse brain. This quality can be represented by *g* and its absence by *G*. When MEL and NWS interact, the recombinants obtained have qualities which can be represented by the scheme

ABDF-CEG + *abdf-ceg* → *ABDF-ceg*.

When the recombinants had been shown to be stable by repeated limit dilution passage, back recombination experiments were undertaken with results of the type shown:

ABDF-ce + *abdf-CE* → *ABDF-CE* + *abdf-ce*
(Lind and Burnet, 1953),

ABDF-ceg + *abdf-CEG* → *abdf-ceg*
(Lind and Burnet, 1954).

It seems very clear from this that the qualities *ABDF* are relatively closely linked and that *CEG* form another linkage group. The simplest interpretation of the facts so far given might be that two chromosomes or, as we prefer, two gene threads are concerned. The objection to this, however, is that there are other ways in which the characters can be recombined. When CAM, an A′ strain, is crossed with WSE the recombinants obtained were

$$A'bDFCE + abdf\text{-}ce \rightarrow A'bD\text{-}fce + abd\text{-}FCE.$$

This suggests that apparently WSE can much more rarely give linkage groups *abd-fce* instead of *abdf-ce*. Then a number of examples of simple transfer of *g* (neurotropic quality) have been observed, for example,

$$aB'df\text{-}cEg + abdf\text{-}ceG \rightarrow aB'dfcE\text{-}G + abdfce\text{-}g$$

(Lind and Burnet, 1954).

We have, therefore, preferred to believe that all these qualities are carried in a single gene thread which can break at various points in the order of frequency 1, 2, 3,

A	B	D	F	C	E	G

3 (between D and F), 1 (between F and C), 2 (between E and G)

and that when appropriate linkage groups, identical or allelic, are available new gene thread sequences can be taken up. As in the discussion of heterozygosis in the previous section, we must also here lay down the rule that viruses heterozygous for serology (AaA') are genetically unstable, so that when there are two or more types of *ABDF* linkage groups available the eventual descendants will show only one of them. To what extent a similar rule holds for the *CEG* group is difficult to say. From the findings in regard to the change from phase O to D which essentially represents a minor change in the quality *C* we should expect that the requirements are by no means as strict and that in many 'pure clones' there may be considerable heterozygosity in regard to the second section of the gene thread. Two of the qualities *E* and *G* are virulence ones which will be discussed in the next section and the O–D results indicate that the *C* group of alleles may have rather similar qualities. Nevertheless, we are impressed with the fact that the overall picture divides the qualities into two well-defined groups and points to the existence of genetic determinants in linkage relationships.

Redistribution of virulence. When a strain of influenza virus has been isolated and, following repeated passage, adapted to grow freely in the allantoic cavity, it will normally show no virulence for mice. If virus of high titre (100 hæmagglutinating doses or more) is inoculated intranasally, some lung consolidation of toxic character results with minimal multiplication of virus. Smaller doses produce no visible lesions. If passage of virus from mouse to mouse is continued using either ground lung emulsion or bronchial washings taken two days after each inoculation, lesions appear about the eighth to tenth passage and increase progressively in intensity. By the twentieth passage the virulence will usually be such that a dose of 0·001 hæmagglutinating doses produces half consolidation of the lungs in seven days. This virulence is maintained when the virus is re-isolated and passaged in the allantoic cavity for a small number of passages at least. In this way it is possible to have two substrains of an influenza virus both readily grown in the allantoic cavity, but one highly virulent for another host, the other quite avirulent. This provides very suitable material for the study of the development and inheritance of virulence.

The essential features of the development of mouse virulence appear to be (i) a variable delay in the appearance of any lesion-producing capacity, and (ii), once such capacity has appeared, a smooth and rapid increase to maximal activity. When single clones are isolated at an early stage of adaptation, many are found to have no virulence for mice, others show a low virulence (Burnet and Lind, 1954 *d*). Adaptation is, therefore, a step-like process with many inheritable grades. It is not a question of a single mutation to a fully virulent form with gradual replacement of the avirulent strain by the virulent. At any given intermediate phase one would probably find representatives of several different stages. The impression we received gave rise to the hypothesis that at a certain stage a 'virulence gene' appeared by mutation and that the observed virulence of any given fluid was a measure of the average number of virulence genes.

In one way or another the virulence genes increased in number faster than other aspects of the genome and were distributed at random amongst the virus particles emerging from the replicating pool of each generation. If the maximal numbers of virulence genes that a particle could carry should be 10 and the average replicating pool per cell gives rise to 100 virus particles, we can picture the process as a progressive increase from one to more than a thousand virulence genes per pool with a binomial distribution of genes amongst the particles.

It was also characteristic of the intermediate fluids obtained that on one or two limit-dilution passages in the allantoic cavity the average virulence fell.

When recombination experiments are made, there is a redistribution of virulence. An unpublished experiment in which CAM (original non-mouse pathogenic) and CAM-MP (highly mouse pathogenic derivative) were used gave the results shown in the table.

Redistribution of virulence on recombination

Source	Number of pure clone fluids with mouse pathogenicity				
	O 0–trace	I 0·1–0·5	II 0·6–1·5	III 1·6–3·9	IV 4·0+
CAM	9	—	—	—	—
CAM-MP	—	—	—	—	9
CAM/CAM-MP	4	3	3	4	1

Grades are based on average lung lesions of mice inoculated with 1 A.D.

Many other experiments of this general type have been carried out and a general account of the results is given by Burnet and Lind (1954 *c*). For obvious reasons it is desirable for recombination studies that the virulent and avirulent strains being crossed should also be differentiated by another marker, preferably a serological one. Most of the experiments have been concerned with the redistribution, in the progeny, of pathogenicity for the mouse lung or of capacity to produce fatal infection on intracerebral infection in mice. Most of the results were summarized by Burnet and Lind (1954 *c*), but the following gives an outline of their general character.

A. *Mouse lung virulence.* Free redistribution is seen in the system LEE B/MIL B with an almost symmetrical gain of mouse virulence by MIL-type recombinants and loss by LEE-type recombinants (Perry and Burnet, 1953; Perry, van den Ende and Burnet, 1954). In other combinations it is usual to find a more or less extensive loss of virulence by the 'donor' strain with much less evidence of gain in virulence by the initially avirulent strain. In the MEL/WSE and WS−/M+ systems there is virtually no redistribution of mouse lung virulence except for the appearance of some less virulent strains of the donor type. When CAM or PERS is crossed with WSE a proportion of virulent recombinants of CAM or PERS serology is obtained, but the intensity of the virulence is low and variable (Burnet and Lind, 1955).

B. *Pathogenicity for the mouse brain.* Here a strikingly wide range in the degree of virulence as shown by survival rate and time from inoculation to death has been found in both sets of experiments, NWS/MEL (Burnet and Edney, 1951) and NWS/WSM etc. (Lind and Burnet, 1954).

The greater ease with which virulence was lost in these experiments is probably related to the other frequent finding, that when a pure clone of a recombinant is submitted to repeated re-isolation at limit dilution in the allantoic cavity, there is almost regularly a downward drift in virulence. This is not seen when a well-established type like NWS is given similar passage.

The lability of virulence demands some explanation other than the occurrence of a series of mutations and back mutations. There are three important points to be kept in mind in seeking an interpretation.

(1) Each of the virulences with which we are concerned, *Ee* for chick embryo, *Ff* for mouse lung and *Gg* for the mouse brain, has its place in the two linkage groups discussed in the preceding section, *ABDF-CEG*, one in the first, two in the second linkage group.

This means that any concept involving virulence gene dosage must be compatible with the interchange of linkage groups already described.

(2) The ready loss of virulence level on limit-dilution passage in the allantoic cavity suggests a straightforward reduction in the concentration or number of some genetic unit rather than anything resulting from a series of back mutations.

(3) The development of a new virulence is compatible with the initial occurrence of a mutation followed by a non-mutational process that increases the amount of a virulence factor provided by the initial mutation. The possibility of more than one mutation is, of course, not excluded.

In broad terms, the results are only explicable (in terms of genetic conventions) if we assume the existence of multiple genes responsible for virulence and their relatively free redistribution amongst the progeny of a double infection.

The most orthodox suggestion might be to regard each increase in virulence as due to a new gene mutation in a unitary genetic mechanism. Accepting the linkage groups phenomena, there seems to be no way in which such a concept could allow the observed redistribution to occur. In previous discussions (Burnet and Lind, 1954*c*) we have adopted the idea of free virulence genes capable of replication at rates differing from those of the other components of the genome. It was assumed that these virulence genes could be attached to the appropriate linkage groups. The main objection to this hypothesis was the absence of any analogy with other genetic phenomena.

An alternative perhaps equally devoid of analogies to the accepted interpretations of higher genetic systems is to regard virulence as being determined by the number or proportion of gene threads which carry an allele for virulence towards the host under consideration. In round figures we can assume, as before, that there are 10 gene threads on the average in each viable virus particle and that 100 virus particles arise from each intracellular replicating pool. On one thread a mutation of F to f occurs which (i) confers the capacity to initiate damage which, when cumulative, is responsible for the production of lung lesions, (ii) confers a selective advantage for survival of this type of gene thread as compared with others, in the murine host cell, and (iii) may diminish the relative survival value of the new type of gene thread in allantoic cells.

On these assumptions passage in mice would result eventually in the complete replacement of original gene threads by the new form. An experimental test for the validity of this hypothesis could be based on the deduction that, while intermediate phases should show a significant drop in virulence at each re-isolation at limit dilution in the allantoic cavity (which is the experimental finding), fully adapted virus should be stable, a specific back-mutation being required to allow for the possibility of down-grading virulence. This should be tested experimentally.

It should be remembered that differences in the virulence of two viruses may be related either to the process by which infection of the relevant cell is initiated or to some intracellular activity. Some of the difficulties in understanding the genetic behaviour of virulences may arise from the existence of these two mechanisms. There are hints that differences of virulence as between host species are more closely related to surface qualities of the virus, while differences in the intensity of attack on different tissues of the same host are more likely to be concerned with intracellular actions. It seems probable indeed that most of the present apparent incompatibilities between the phenomena we have called interchange of linkage groups and redistribution of virulence will be resolved when the respective contributions by what can be called surface or somatic components on the one hand and intracellular or reproductive ones on the other are properly understood. Work with this objective in view is now in progress.

A good example of the association of a somatic change associated with a striking change in virulence is the change from phase O to D observed in newly isolated influenza A strains. Though logically a discussion of mutational processes should precede rather than follow a section on recombination, it is more convenient to use the concepts derived from the recombination work to interpret the mutational changes rather than vice versa.

8. *Mutation*

The change from phase O to D in influenza A can be taken as a prototype of mutation in influenza virus. Its occurrence is part of the necessary process by which an influenza virus strain is rendered suitable for conventional laboratory work and is, therefore, of high significance for the laboratory worker. Burnet and Bull (1943) were the first to show that when an influenza A strain is first isolated from a human patient, the fluid agglutinates guinea-pig or human red cells but is almost or completely inactive against fowl cells. With one or two passages in the amniotic or allantoic cavities the character of the virus changes so that it agglutinates fowl cells to the same or higher titre than mammalian cells. This is the most striking feature of the change from O phase to D phase, but an equally important difference is the inability of O phase virus to infect the allantoic cavity. Everything indicates that the influenza virus A in its 'wild' form as a human pathogen can survive only in the O phase.

Once multiplication has occurred in the amniotic cavity a proportion of mutants appear which differ (i) in agglutinating fowl cells, usually rather weakly at first, and (ii) in being able to initiate growth in the allantoic cavity. Although most workers concerned with the isolation of influenza viruses agree that first passage fluids agglutinate human or guinea-pig red cells better than fowl and that virus can only be isolated with regularity in the amniotic cavity, there had been no full confirmation of Burnet and Bull (1943) until the appearance of Mogabgab *et al.*'s paper in 1954. They found that if influenza A virus was isolated in tissue cultures of embryonic human lung, the virus produced had the quality of pure O phase virus.

The most important feature of our results (Burnet and Bull, 1943; Burnet and Stone, 1945) was the demonstration that, by the use of limit-dilution passage and certain technical precautions, a strain of influenza A could be maintained for 24 passages in the true O phase. Though it was not fully recognized at the time the experiments were done, this result seems to have some important implications in relation to the process by which specific virus protein is synthesized in the infected cell. The analysis of the enzymic and adsorptive qualities of O and D phase virus by Stone (1951) showed that the O phase virus had neither adsorptive nor enzymic action on fowl cells, that it was not inhibited in the indicator state by ovomucin (a typical mucoid of avian origin) and in the active state had no enzymic action on this substance. By contrast, O phase virus agglutinated human cells, removed receptors from them by enzymic action and in the indicator state was inhibited by a mucoid of human origin (ovarian cyst).

The first change observed to ω̃ or false O phase gave virus with very weak fowl-cell agglutination which, however, was more marked when the mixtures were held in the refrigerator. It could be shown that such virus could make contact with only a very small proportion of the fowl-cell surface receptors and, at room temperature or higher, rapidly destroyed this proportion by enzymic action. A virus fluid of this incomplete O phase, when absorbed with fowl or mouse red cells, showed an approach to the pure O type, and the most satisfactory way of ensuring the maintenance of O on repeated passage was to make the transfers with virus harvested from the chick embryo lung and absorbed with fowl cells.

In the light of current hypotheses of influenza virus genetics, the interpretation of the change from O to D would be as follows:

(1) In the absence of mutation the functional character of the virus is maintained unchanged in an alien environment. This holds at least for the configuration of the adsorptive enzymic mechanism which is an essential part of the machinery of infection and virus survival. An important experiment which we have never found an opportunity to make would be to test whether O phase virus after 20 or more passages in the chick embryo still had normally high infectivity for man. It would obviously be of very great interest to know whether all characteristics were retained by this procedure or whether some were directly modified by the mere fact of replication in chick embryo rather than human respiratory cells.

(2) The published evidence is sufficient to indicate that there is a gradual change from O to D. The change is certainly

not a simple replacement of O virus by an increasing proportion of standard D virus particles. Burnet and Stone (1945) guessed that there might be two intermediates, but stated that they had no evidence to exclude a much larger number. As far as the evidence went, phenotypic and genotypic changes went in parallel, although the necessity that the phenotypic character mainly studied, the fowl/guinea-pig hæmagglutinin titre ratio, must be measured on a large population of individuals, presented an almost insuperable obstacle to effective analysis of this point.

(3) The simplest interpretation compatible with the facts, not of course necessarily a correct one, is that only one mutational change is needed—one which results in the appearance of a surface enzymic-adsorptive protein molecule that can react with mucoid receptors of avian pattern as well as of human patterns. If we assume that there is a genetic structure based on a number (between five and fifty) of equivalent gene threads in each infective particle, the process takes on the character that we have discussed extensively in relation to the development of virulence. As soon as a mutation in a single gene thread occurs, a proportion of new surface units will appear and any virus particle in which they are incorporated will have a survival advantage in the chick. The possibility must also be considered that the mutation confers a direct advantage in regard to intracellular replication of the mutated gene thread compared to its unchanged fellows. If we assume that there are ten gene threads and 200 corresponding functional molecules on the virus surface, then we might picture a cell that is infected with a particle carrying one mutated thread, giving rise to particles in the next cell generation with an average of one mutant gene thread and of 20 corresponding surface molecules. However, any type of random distribution will give a few particles with three and two gene threads instead of one. Some of these

may also have something more than 20 'new' molecules on their surface and will have a normal chance of inducing infection in the next cell generation, while their progeny will have a sharply increased chance of survival. In this way there must be a steady increase in the average proportion of mutant gene threads per particle and of the proportion of 'new' surface units. There is no way of excluding the possibility that more than one type of mutation is concerned. The regularity and speed of the process point, however, very strongly to the occurrence of a single type of mutation rather than to a variety of random alternatives.

But details are unimportant. The crux of the matter is that the genetic character of the virus is dependent only on what it brings into the cell. Change occurs essentially by the same process of mutation and selective survival that is at work in higher forms.

The point to be stressed in the present connexion is the necessary deduction that any use which is made of the protein synthetic mechanisms of the host cell does not appear to influence in any recognizable fashion the specific patterns of the virus macromolecules.

9. *Summary*

The availability of the chick embryo and the hæmagglutinating power of influenza viruses have allowed a massive elaboration of somewhat artificial experiments. Thoroughly domesticated strains of virus have been used with only a remote resemblance in many of their qualities to the human pathogens from which they derive. Nevertheless, it seems legitimate to claim that only by the utilization of this highly artificial system has it been possible to gain an insight into the nature of the virus and its multiplication, which is as relevant to pathogenic natural strains as to their laboratory derivatives.

The interpretation that has been presented and which is summarized in the diagram, Fig. 6, owes a great deal to Hoyle, but will probably not find favour in the eyes of other virologists interested in influenza viruses. Their objection to Hoyle's view, that it did not provide any interpretation of the highly individual specific properties of the virus, will probably extend to the present picture. The suggestion that what is contained in the virus particle is merely a portion of the 'multiplying pool' that has been trapped by the formation of a surface, has provoked an instinctive disagreement in several with whom the idea has been discussed.

Nevertheless, an attempt has been made to discuss, in terms of the model, all the phenomena involving influenza viruses that have been adequately studied experimentally. The only significant exception is the result of research on the inhibition of multiplication of influenza virus effected by substances of small molecular weight, in which some important recent work is associated with the names of Ackermann, Tamm and Horsfall and Eaton. As will be discussed in the concluding chapter, such studies, although they may have very important practical results, can make virtually no contribution to the understanding of phenomena based on the interaction of macromolecular pattern. They may eventually become highly relevant to the processes of intermediary metabolism in the infected cell by which the necessary building blocks are made available to the multiplying pool, but this is not relevant to the picture at the level with which we are concerned.

The wide range of morphological, chemical and functional data which can be covered by the formulation can never be taken as a proof of its correctness, but is at least an indication of current usefulness. I can but stress particularly that it is not easy to see any alternative hypothesis which will account for three sets of phenomena:

(1) the existence of grades of incompleteness, including viruses capable of a single generation only of replication;

(2) the existence of RNA only in the virus;

(3) the three types of genetic interaction that can occur.

All three have been the subject of close study in the Hall Institute and the hypothesis presented has grown out of the efforts made to interpret them.

In view of the fact that very recently independent confirmation of Ada's finding that the influenza virus contains no DNA has been forthcoming (Liu *et al.* 1955), it is worth underlining that the influenza virus is as yet the only well-studied genetic system in which DNA plays no part. Specific pattern is here carried by a system of which protein and RNA seem to be the only significant components. This in itself is an important addition to the developing concepts of protein biosynthesis which endow RNA with an essential template function. An attempt has, therefore, been made to describe the process of virus multiplication in basically the same terms as have been used in regard to the protein-synthesizing processes by which adaptive enzymes or antibodies are produced.

CHAPTER V

THE SCOPE OF BIOLOGICAL GENERALIZATION

The impulse to prepare this monograph arose primarily from an attempt to understand what was happening when influenza viruses multiplied and recombined in the host cell. This necessarily involves a primary consideration of the mutual interrelationship of protein and ribonucleic acid, the two substances which alone seem to be involved in the smaller viruses and which are almost certainly the significant and specific components of influenza virus. The question of protein synthesis and the part ascribed by many writers to RNA in that process is clearly vital. This discussion has, therefore, grown into something which has so far been concerned in some detail with three examples of the production by cells of protein with 'abnormal' pattern and function. The aim that emerged in the process was to find either some general description of protein biosynthesis applicable to all three fields and, by implication, to *all* protein biosynthesis or to try to envisage what experimental approaches in these fields might favour the eventual production of such generalizations.

In attempting any such discussion it was impossible to avoid some consideration of the even more basic question of what was the real objective of biological investigation at the academic and theoretical level. What was its human and social justification on the one hand? What were the approaches that would lead to the type of understanding desired, on the other?

There is an infinite field for study amongst living organisms. Any worker of intelligence and ingenuity can take the most unpromising-looking organism and find a life-time's work. If he has the qualities of a leader in science he will almost inevitably build up a school of investigation that will allow a dozen younger men to produce work of first-rate quality on one aspect or another of the organism chosen. One has only to mention those completely 'useless' organisms *Drosophila*, *Paramecium*, *Neurospora* and phage T2 to illustrate this. Since it is characteristic of all the natural sciences that successful work always opens up opportunities for new advance—perhaps even more in biology than in the physical sciences—one can see ahead an infinite elaboration of biological investigation limited only by the community's allocation of the funds and leisure required. It is arguable that this is a thoroughly desirable prospect, that if those with the ability and liking for biological investigation can be occupied with satisfying work, it is quite immaterial whether the topic chosen has any relevance to anything else or not. The only qualification is that the investigation must be carried out with full regard to scientific integrity.

At the present time it is not quite possible to accept this point of view and, looked at realistically, there seem to be two main reasons for the types of research actually undertaken in present-day biology. In any particular field of investigation there will usually be some practical objective towards which an understanding of the phenomena may be expected to contribute. In most cases this will be at least the nominal reason why funds are made available for the extension of such research. Work on influenza virus genetics, for instance, may be justified by the hope that it will eventually allow the production at will of virus strains with desired combinations of properties and that conceivably it could provide the clue to the interpretation and control of pandemic outbreaks of lethal influenza. Reasons of this sort, however, have often little cogency for the typical academically minded investigator who finds interest in such topics.

To him the immediate objective is probably to find a field in which he can carry out a craftsman-like performance which will be applauded by his fellow-craftsmen and help to consolidate or improve his position in the social hierarchy of science. This immediately raises the problem, what is acceptable academic research in biology, using 'acceptable' in the pragmatic sense of actually improving the worker's status amongst his fellows. The requirements seem to be as follows:

(1) The results obtained shall be new, but not so novel that they cannot be readily related to the results of previous work in the same general field.

(2) The work shall be done efficiently, adequate technical and logical (including mathematical) methods shall be used, with sufficient description to ensure that the work can be repeated by any colleagues with the necessary facilities and skills. Except on rare occasions, recognition will not follow until the results have been shown to be reproducible in another laboratory.

(3) The results shall have an internal consistency which will allow at least some measure of generalization. This might range from the statement that such and such a virus contained $0{\cdot}8\% \pm 0{\cdot}1\%$ of RNA to a complete account—to take a hypothetical example—of the natural history and distribution of serum hepatitis virus in human beings.

(4) Generalizations are acceptable in proportion to (i) their applicability to some practical problem, (ii) the width of the academic field or fields to which they can be applied, particularly when they open up new *experimental* approaches in other fields, (iii) their capacity to be incorporated into the body of general scientific knowledge as presented in the standard senior text-books of the period.

The really great generalizations of biology—the theory of evolution by natural selection, Mendelian inheritance, the linear arrangement of genes, the uniformity of glucose metabolism, immunization by attenuated pathogens, pituitary control of endocrine functions and the like—are rather outside the normal process of investigation. They are always dependent on the work of many groups and, in the modern period at least, they develop gradually at first in incomplete and often unbalanced form, usually taking their 'text-book' form when an outstanding book or review by a leading worker appears. It is evident in recent biology that major generalizations do not often emerge, nor, when they are sensed, is it easy to provide the evidence that is needed before they are acceptable. At the present time the incipient generalization of the widest biological interest is probably the one with which this monograph is mainly concerned—that protein is synthesized by or on a RNA+ template. As will have been evident, it has proved very difficult to obtain more than indirect evidence in favour of the conception.

Without attempting to analyse the factors, other than the acceptability under the existing traditions of scientific work, which give significance to a generalization, there is still scope for a limited examination of the difficulties that are evident in regard to our particular problem of protein biosynthesis. One important possibility is that we are here reaching a field where the three types of convention so far used to describe biological happenings all fail to provide the concepts that are needed. (i) In the first place we have the morphological approach to anatomy, histology and elementary physiology. This is now being refined towards its limits by the use of the electron microscope and the development of histochemical methods, but it is still essentially an extension of direct sensory observation of biological material. (ii) At the level of evolution, genetics and population genetics we can use logical and mathematical methods derived again fairly directly from normal human experience of animal breeding, various aspects of demography, games of chance, etc. (iii) Finally we have the direct experimental approach in the tradition of laboratory investigations in chemistry and physics, the characteristic examples at the present time being biochemistry and electrophysiology. The normal approach to protein biosynthesis would be almost wholly along the third path of biochemistry and physical chemistry. Any discussion of the limitations of such an approach will naturally find its examples in the field of biochemistry.

The biochemists now so dominate the picture in biology that it may seem foolhardy and churlish to depreciate the success that has been obtained in dealing with very many aspects of the chemistry of living substance. But in a sense all these successes have been at a superficial level. The elucidation of the process by which glucose is oxidized has been a masterpiece of experiment and logic, but it leaves us in deep ignorance of the structural basis and the functional co-ordination of the processes that have been uncovered, and even further from an adequate understanding of the manner in which the organization is laid down in inheritance and development. The more optimistic biologists, while agreeing with this as an expression of the present situation, would probably consider that we are already well on the way at least towards diminishing the gap between the morphological and the biochemical approaches. A direct attack on cell structure by electron microscopy is beginning; the difficulties of producing thin enough sections have been solved; but so far there has been little advance in the interpretation of the appearances found. Methods of differential 'staining' have yet to be developed. Considerable progress can undoubtedly be looked for in the next few years, perhaps particularly in the comparative study of cell types highly specialized for certain functions. One can notice already a concentration of interest on the secreting cells of the mouse pancreas and the chloroplasts of plant cells for just this reason. Sections of the exocrine cells of the pancreas show an elaborate arrangement of membranes or flattened sacs which can hardly have any other function than protein synthesis. The appearance of somewhat similar pairs of lamellæ in mitochondria and chloroplasts is already leading to speculation about the relationship of these structures to the corresponding functions.

It must be stressed, however, that there is a great difference between developing a reasonable picture of some biological function and being able to derive from such a generalization methods for the control of the phenomena at either experimental or practical levels. Suppose, for instance, that it should emerge that the double lamellæ in the cytoplasm represent 'production lines' between which metabolites are moved past enzymes ranged in an appropriate series to carry out some sequence of chemical action. It might even become possible to obtain a genetic variant in which a block exists at some stage of the chosen sequence and in which a morphological change in the ultramicro-structure can be correlated with the block. The effective application of such knowledge, however, is likely to be small. It is virtually inconceivable that it would in any reasonably direct sense allow the modification at will of the metabolism of the cell concerned beyond what could be accomplished by genetic and enzymological work alone. The attempt to press the structural, physical and chemical approach to the understanding of living process seems to have reached the phase of diminishing return for the effort involved. We are approaching an asymptotic barrier and it may be that some modification in the outlook and approach of theoretical biology will soon be needed.

1. *Information theory in biology*

It seems that this dilemma is not confined to those aspects of biology with which we are directly concerned. A similar situation probably exists where electrophysiology impinges on behavioural problems or ionic physiology on homœostasis. As J. Z. Young (1954) points out, these are the fields where we have so far failed to find 'appropriate models that shall provide a language with which we can speak about these complicated processes'. Young was referring largely to the difficulty of formulating a scientific approach to the problem of memory, but the remark, as he recognized, is equally applicable to the domain in which genetic qualities are manifested in the appearance and functioning of organic macromolecules.

In this search for a language that will function in these twilight zones where chemical description of molecular structure and function breaks down, much interest has been taken in the development of modern communications theory. Since 1945 there has been a widespread recognition amongst scientists and the public generally of the importance of the principles which have emerged from experience in the development of electronic communications and control systems. In the physical field this has led to the production and operation of an increasing variety of devices for the transmission and analysis of information and for the automatic control of 'purposive' processes like the functioning of a chemical plant or the flight of a guided missile. Amongst biologists, the neurologists have shown the greatest interest in the principles involved and it seems that the designers of computers and control systems are developing a reciprocal interest in the concepts of neurophysiology. In our field, the relation is a good deal more distant, but one cannot escape the attraction of the general approach.

This monograph was originally conceived as an attempt to develop something analogous to a communications theory that would be applicable to the concepts of general biology. However, it has not been found possible to make any serious use of the already extensively developed concepts of information theory in the strict sense. In part this is due to the weakness of the writer's training in mathematics and his lack of interest in mathematical logic. A much more important reason, however, is that only the most generalized sketch of an outline has yet been given of how information theory at the strict level can be applied to biology.

The only extended account of such an approach that I have been able to find is the symposium edited by Quastler (1953). This gives a clear indication (*a*) of the fact that biological organization can only be understood quantitatively in terms and conventions that are concerned with ideas of the quality of order, regularity, specificity, certainty and the like, i.e. in terms of information theory not of classical thermodynamics, and (*b*) that the approach is likely to be one of extreme difficulty due simply to the complexity and hence the very large content of 'information' in any living organism.

As the editors point out in their introduction, the gap between a simple and a complex organism, e.g. a bacterial cell and man, is small in comparison to the gap which separates the simplest living things from the most complicated non-living systems, a giant 'electronic brain', for example.

The basis on which attempts to apply information theory to biological processes have been developed is clearly expressed in Schrödinger's (1944) popular statement on *What is Life?* In his words an organism must feed on negative entropy or more strictly it can only exist by freeing itself from all the entropy it cannot help producing while alive. Negative entropy is a measure of order, and information can be equated to $-\alpha S$, where S is the entropy of the system and α a constant, when appropriate units are used. In general, information is measured in terms of the number of binary choices, 'bits', needed to define the system in question. It can, therefore, be brought into some sort of quantitative relationship to any system in which the thermodynamic situation is understood.

In another direction interesting quantitative analogies can be drawn between the distribution of various numbers of some twenty types of amino acid residues in a protein molecule and the distribution of twenty-six letters in a paragraph of English. In a certain sense it is reasonable to think of the biological function of the protein as broadly analogous to the meaning of a paragraph. But 'meaning' in this sense is not equivalent to 'information'. Little would be gained by studying a paragraph in ways strictly analogous to those applicable to a protein. It would be hopeless to seek its meaning by analysing a paragraph in terms of the number of times *a*, *b*, *c*, etc., appear or even by finding a variety of rules of sequence, that *u* always follows *q*, that no sequence of more than two of the same letter ever occurs and so on. The meaning of a paragraph can only be found by what is mathematically a vastly more subtle process. Yet that meaning is immediately apparent to any literate English-speaker by what without too much violence can be regarded as a biological process analogous to the 'recognition' of a functional protein based on some form of specific complementary pattern.

Schrödinger puts the position very clearly by showing first that the potential complexity of nucleic acid or protein molecules was such as to make it conceivable 'that the miniature code [in the chromosomes] should be in one to one correspondence with a highly complicated and specified plan of development and should somehow contain the means of putting it into operation'; but also that 'no detailed information . . . can emerge from so general a description'. He feels rather that the development of biology will demand the emergence of 'other laws of physics hitherto unknown' which, once revealed, will form an integral part of physical science complementary to the laws of classical physics and thermodynamics. These laws must be such as will allow the derivation of 'order from order', in a system where at the same time normal entropy-increasing processes are constantly going on.

Such views, however, are mainly relevant to the problems of intermediary metabolism and the interpretation of the organism at the thermodynamic level. We have been concerned with matters which allow us to take the metabolic background for granted—much in the fashion that a communications engineer can take the source of his current and the nature of the telephone wires he is using as something necessary but quite without interest for the problems with which he is specifically concerned. It may be more useful to follow the analogy with the communications engineer's specific problems.

In any transmission of information we need a channel by which the information is transmitted and a series of interrelated codes in which any required message (or instruction) can be expressed. Speech is itself a coded arrangement of sounds in air; by an appropriate transmitter it can be alternatively coded in the form of a pattern in radio waves and reconverted to sound or to another type of pattern on magnetized tape in a corresponding receiver. There may even be some biological analogies with the finding that in all these recordings there tends to appear some distortion of the original pattern and a background of 'noise', i.e. meaningless physical action of the same general quality as the physical action carrying the meaningful message.

Genetic experiments may tell us that the difference between two antigens is related to a single gene difference. Physical and chemical study may suggest that the two protein antigens differ because they contain amino acid residues arranged in two different sequences. Current theory, not too confidently, places the hereditary code in the DNA macromolecules of the nucleus. The first problem in the development of a communications theory of the cell is to indicate the form taken by the 'channel' that leads from the code carried by the genes to its manifestation in the chemical structure of an antigen or of any other genetically determined component of the cell.

We can be fairly confident in claiming that macromolecular pattern is the basis of the flow of information in the cell in the same sort of sense that the printed word is the basis of the flow of information from one laboratory to another. For an effective channel to function on such a basis there are four requirements.

(1) Means by which specific pattern can be replicated in the same medium (medium being used to describe the general chemical type of the macromolecule concerned—protein, nucleic acid, polysaccharide or lipid).

(2) Means by which a meaningful pattern can be recoded in another medium. In some way a DNA molecule can convey information that causes a pneumococcus to produce a new specific pattern in polysaccharide.

(3) Means by which specific pattern can be recognized as such and convey its 'instructions'.

(4) Means by which more than one channel of information may converge or combine to affect a common target. There is some evidence that a single antigen may be influenced by more than one gene.

The likely clue to the way in which these requirements have been fulfilled lies in the existence of complementary or, perhaps even more important, nearly complementary pattern. There must be something about the way in which proteins are built that allows them to be potentially capable of making specific complementary pattern (SCP) relationships to the significant configurations (groupings) of *all* the organic molecules and macromolecules that make up living substance. It is axiomatic that only substances susceptible to enzymic breakdown can occur in living cells. One feels that there may be a noteworthy generalization awaiting the organic chemist who can show that the standard 24 amino acids represent an alphabet which, by appropriate mutual arrangements, can provide SCP's for *all* the configurations that are possible in biologically acceptable molecules. Thinking along the same lines, we might hazard the further guess that another chemist might find that there were, in fact, two types of SCP relationships that could be provided. The difference between the two would determine the difference between enzyme-substrate reaction, on the one hand, and antigen-antibody union, and perhaps ⌜enzyme, competitive inhibitor⌝ union on the other. In these SCP relationships we have clearly a means by which information can be conveyed in a highly selective manner; it is like a 'scrambled' radio message which is meaningless to all receivers except the one with the pattern of action that will 'unscramble' the message.

The next requirement of a biological communications system of this type is a means by which pattern (code) can be imprinted during synthesis of the macromolecular medium concerned. As an extension of the concept enunciated in the last paragraph, it seems biologically necessary to endow protein in the appropriate environment, with the capacity to act as a template which will guide the synthesis of significant macromolecules in another medium. In the general body of this monograph much use has been made of the hypothesis that new RNA is synthesized against a protein template, but it seems that we must also make a similar demand for the synthesis of complex polysaccharide or phospholipid against protein (or sections of protein) carrying SCP of appropriate type.

The position of DNA in the system is clearly of crucial importance. At the present time the twin helix formulation of Watson and Crick is regarded as pointing almost peremptorily to a mechanism of direct replication that looks as if it were 'designed' to make any deviation from exact replication impossible. This would fit well with other indications of the stability of DNA which might almost be regarded as the cell's repository of the master keys, the ultimate standards to which all the working jigs and patterns must conform but

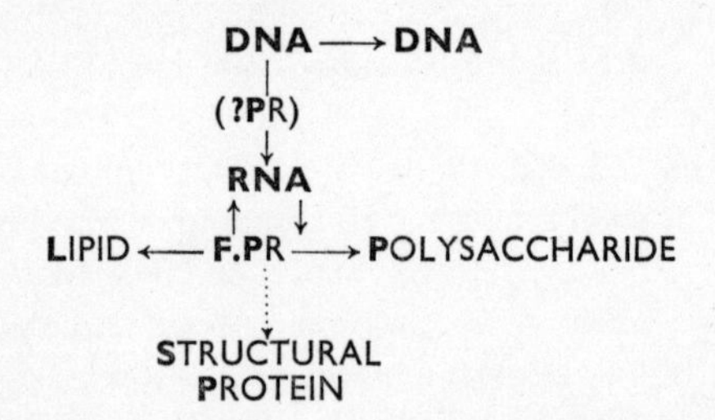

FIG. 8. A diagram to illustrate a concept of the flow of specific pattern in the cell. Each arrow indicates that the pattern on the medium from which it starts acts as a template for the medium to which it points.

which are too precious for ordinary workshop use. It would follow that the coded instructions of DNA must in some way be transferable to RNA in the nucleus and that perhaps at one or two removes the pattern at length reaches the cytoplasm where the final functional protein is produced. The relation may be as expressed in a diagram (Fig. 8) in which an arrow signifies that the pattern on the medium from which it starts acts as the template for the medium to which it points.

Functional protein (F.Pr) is differentiated in the diagram, perhaps on inadequate grounds, from structural protein such as collagen. It should be stressed that the general principle of the possibility of transferring coded information in pattern on one medium to a different coding of pattern on another medium is what is important, not the detail of the actual sequence in the cell. For example, we should have no quarrel whatever with anyone who claimed that the transfer of pattern from DNA to RNA was mediated through nuclear protein.

On this hypothesis, DNA pattern is the only one which is replicated directly; all the others make use of a SCP relationship. As we have mentioned earlier, this provides a versatile and effective means of rapidly increasing the amount of any patterned material that is required.

One implication of the hypothesis that is particularly important in relation to the idea of genocopy, used in discussing antibody production, should be mentioned. If protein can potentially take on a SCP relation to any biologically significant molecule and if functional protein is always synthesized on a RNA+ template, then we must postulate that RNA can carry all the codes which can be expressed in protein pattern. This makes it legitimate to assume that if a RNA+ template is distorted by the incorporation of a foreign antigenic determinant so that it produces a new SCP in the protein synthesized, then whatever the character of the new protein it can induce the synthesis of a SCP pattern to correspond in RNA. This then becomes the genocopy.

2. *The application of pattern concepts to biological problems*

The approach to biology by the use of the concept of specific complementary pattern has been more or less consciously developing since the time when Ehrlich first put forward his ideas on immunological specificity. In this final section an attempt may be made to indicate the way in which the concepts of 'communication through macromolecular pattern' are nowadays amongst the normal tools of every biological investigator and theorist.

In the first place, it is very evident that in fact much more use is made of SCP relationships than of more physico-chemical concepts in most areas of experimental biology. It is probable that the larger bacterial viruses of which T2 is the type have been more intensely investigated at the fundamental physico-chemical level than any other organism. Yet it is hardly an overstatement to say that the results obtained in ten years' intensive study have virtually no relevance for any other type of organism. It has not even thrown much light on the metabolism of the host cell and it has had no influence whatever on any more practical human affairs.

It may be illuminating to consider those aspects of bacteriophage research which have been of value to something other than the understanding of the system actually under study. They are almost all concerned with specific relationships between virus and host and not at all with the process by which multiplication of virus takes place. At the academic level of experimental biology, the most important use of phage has been as a highly selective inhibitory agent which can be very usefully applied to the study of bacterial mutation. The only practical applications of bacterial viruses to medicine and industry are similarly based. Type differentiation amongst typhoid bacilli and staphylococci particularly have been very valuable for epidemiological investigation. Boyd (1950) has suggested that similar use could be made of the lysogenic phages carried by many salmonellas as identifying marks in epidemiological work. Industrially, the cheese industry has overcome most of its difficulties with phage action on streptococcus 'starter' cultures by having a wide range of different resistant mutants available to replace any culture that is rendered useless by the entry of an active phage.

This general character seems to be typical of all the fields of biological research. The 'useful' approaches are concerned with arranging various series of reactions based on SCP relationships into a logical pattern that can be effectively used for further experimental or practical purposes. We have been concerned with three major topics—adaptive enzyme formation, antibody production and virus genetics—and it is reasonable to take these as representative of other academic biological fields. What are the aspects that have been useful in the broad sense?

(1) Adaptive enzymes have been useful for experimental work primarily because, by their functional titration, they allow a study of the process of new protein synthesis under various conditions. It is their SCP relationship to substrate and inducer that makes them valuable. At the practical level they have a minor value in the biochemical classification of strains of micro-organisms and there are occasions when appropriate use of an inducer will greatly increase the yield of some needed enzyme.

(2) Antibody production is of the greatest importance to medicine and there is a very large industry employed in its exploitation in one form or another. SCP relationships govern the whole activity in the field with the exception of the non-specific methods of concentrating from immune sera the γ-globulin that carries the antibody. The usefulness of an influenza vaccine depends simply on the closeness of the SCP relationship between the antibody produced by the vaccine and the antigenic pattern of the virus responsible for the next epidemic. Virtually all procedures for the etiological diagnosis of infectious disease are finalized by an immunological identification of the micro-organism. And when immunological methods are used as tools in biochemical or other investigations, it is always by an exploitation of the SCP relationship to identify or to remove some component of the system.

(3) The practical side of virus disease has so far hardly been concerned with fundamentals. Prevention has been a matter either of immunological work or of very matter-of-fact interference with the process by which it reaches susceptible subjects—mosquito control, disinfection of fæces, quarantine regulations and the like. Virus genetics has so far found practical use only in the empirical production of avirulent variants for immunization. The success of myxomatosis in at least temporarily controlling rabbits in Australia suggests that in the future there may be legitimate uses to be made of hybrid viruses designed like hybrid cereals for specified uses in pest control. If such work is to succeed, it will depend primarily on our being able to produce the same sort of formulation of the genetic potentialities of specific strains and their progeny as is available in the form of chromosome maps or otherwise for all the organisms that have been successfully studied by geneticists.

This admittedly is an overstatement of the futility of the physico-chemical approach to biological problems. But the expression of such views may serve as a useful stimulus to the more conscious development and use of concepts of specific complementary pattern as the essential means by which the living cell is organized and as the key to the control and modification of the significant biological phenomena.

It would be logical, therefore, to conclude with an attempt to indicate in what directions a more conscious appreciation of the importance of SCP relationship would direct basic biological research. If we accept the view that understanding is only of significance in so far as it furnishes us with means for the control of the phenomenon in question, a fairly definite formulation is possible. Perhaps it does not greatly detract from that formulation to find that it does little more than express in general terms the main types of biological and biochemical investigations currently being undertaken.

In any biological system considered at the level with which we are concerned, the type of control desired will be to modify some end result of a chain of communication. Typical examples are to produce maize or wheat with a desired combination of qualities, to render children immune to poliomyelitis, to remedy the effect of the genetic abnormality in hæmophilia or to prevent the occurrence of leucæmia.

Modification of a communications system of the general type that we are envisaging will require (i) a knowledge of the sequence of carriers, at least in some critical portion of the chain, (ii) a knowledge of the physical and chemical conditions necessary for the transfer of the instructions from one carrier to the next, (iii) a knowledge of the possibilities by which new patterns (from other biological materials) can be inserted at some point into the system so as to switch the process to a new end point, (iv) an empirical exploration of the potentialities of modifying the end product by the introduction of small molecules which may disturb pattern relations at some point.

A few comments may be made on each of these approaches.

(1) The sequence of carriers has provided the central theme of this monograph. There is much to suggest that all organisms with a genetic apparatus replicating by mitosis will be found to have a standard sequence of carriers, at least in the initial stages of the process. The elucidation of that sequence, however, is bound to be difficult; current experience indicates that it will be extremely difficult to obtain direct evidence. In all probability the sequence will have to be deduced from the results of experiments directed towards obtaining data on points (2), (3) and (4) below.

(2) The simplest type of 'transfer' of instructions is the action of enzyme on substrate. The dependence of enzymic action on the concentration of substrate, temperature, *p*H and other ionic requirements makes it highly probable that

other SCP unions will also be influenced by the ionic and other conditions and to this extent will be susceptible to experimental modification. Since most of the reactions involved take place in a highly buffered homœostatic system, it may be difficult to produce a desired alteration in the conditions without provoking many unwanted secondary effects. Where 'transfer of instructions' involves the synthesis of a new carrier, the necessary conditions will presumably be those which we can express crudely as 'the presence of a pool of amino acids or other building blocks and the necessary supply of energy'. A considerable proportion of current experimental work is concerned with attempts to modify the conditions in this pool.

(3) The insertion of a new pattern with resulting modification of the end result is the most direct approach to the field and the two sections on antibody production and virus replication have been written largely from this point of view. Much work is being done on the transfer, by something less than sexual fusion, of genetic characters from one bacterium to another. The classic example is the pneumococcal transforming principle of Avery which has more recently been shown to have a much wider capacity than the conversion of one serological type to another. The evidence still seems to hold that the transforming agent is a DNA macromolecule which can in some way be 'slipped into the pack' of the normal genetic determinants of the recipient pneumococcus. The transduction phenomenon of Zinder and Lederberg (1952) appears to be basically similar, the main difference being that the pneumococcal agent can enter the recipient cell without assistance, while the transferred genetic character in a transduction reaction is carried by a temperate bacterial virus into the substance of the recipient. Of the same general quality is the Berry-Dedrick phenomenon by which fibroma virus can be transformed to myxoma by an

extract of heat-inactivated myxoma material. The results, however, have not been sufficiently reproducible to allow systematic work on the nature of the active agent.

(4) Chemical mutagenesis is the outstanding example of relevant current work and there is reasonable certainty that irradiation-induced mutation must also be regarded as essentially chemical in quality. However, in this field we come immediately on the characteristic feature which, in one way or another, is stressed throughout this monograph—the randomness of all our attempted modifications of biological pattern by the direct approach. A very wide range of dissimilar substances, from nitrogen mustard to aromatic hydrocarbons and manganous salts, can provoke mutations in bacteria. The mutations, however, have no significant relationship to the quality of the stimulus used. The results are about equivalent to those which one would obtain by carrying out an experiment of dropping a variety of missiles each on a series of a hundred typewriters, and assessing the effect in terms of success in typing the alphabet on each machine. Quite a lot of information could, in fact, be deduced about the structure of a typewriter by studying the scripts so obtained in relation to the missile dropped, but hardly sufficient to understand its mechanism, or the process by which it is built.

The possibilities of more direct incorporation of small molecular configurations into natural patterns were exploited intensively by Landsteiner and his successors in their work on semi-synthetic antigens. The study of carcinogenesis, which Green's work is bringing into direct relation with immunological phenomena, is probably leading in a somewhat similar direction. An intermediate condition of high potential productivity is the immunological sensitization of men or guinea-pigs by simple chemical substances which can react with body proteins to confer a new specificity.

The very active field of investigation in which unacceptable analogues of key metabolites are introduced into biological systems can be considered here, although, since the usual object is to prevent the occurrence of a specific process rather than to modify its end point, it might be better to take it under a fifth and final heading. On the basis of the hypotheses that we have been using, the objective is to vitiate one of the synthetic reactions by providing an artificial component which, by taking the place of the normal component (amino acid, purine base, etc.), can distort a pattern sufficiently to render it non-functional. On the whole such work has been more successful in providing partial explanations of empirically discovered actions of drugs than of laying down paths for the biological development of practically effective substances. At the moment it is virtually the only chemotherapeutic approach to cancer and to the extent that the metabolism of the cancer cell resembles that of the normal cell of the host one has to be pessimistic about the outcome. Any effects obtained to date seem to be little more than indications that it is somewhat easier to damage a rapidly growing cell than a more quiescent one. Nevertheless, while interest continues, a great variety of organic chemicals will be tested experimentally and empirically. A positive effect may emerge which, as in the past, can open up a new approach that may yield results of the greatest practical importance. It is, however, almost a summary of the whole thesis of this monograph to say that any such success will remain on an empirical basis, because the macromolecular patterns that are being distorted are beyond standard physico-chemical description and manipulation.

However unlikely it may be that the practical control of biological phenomena will ever make use of the strict concepts of structural chemistry and mathematical logic, we can still be certain that the search for valid generalizations will continue along such lines. I believe that most biologists would accept Schrödinger's contention that there are new laws of physics eventually to be uncovered by the study—necessarily along physical lines—of living organisms. If these are to be sought, they will only be found if biological systems can be devised that are simple enough to allow a thorough physical analysis. One might imagine the possibility that some simple cyclic peptide like oxytocin or tyrothricin could be replicated by a definable system of enzymes in the test tube, all the necessary building blocks, energy, sources, etc., being supplied. It is conceivable—perhaps—that a full description of the process in terms of structural chemistry and thermodynamics might be obtainable. From such data the essential physical laws of a biologically replicating system might be derived. Such a feat is at present virtually unthinkable. If it is ever achieved, it will be by the use of a very large and very complex array of workers, instruments and computing machines beyond anything now available for non-military research. And when the knowledge has been gained it will have very little relevance to any human uses of the peptide concerned and none to those of any other type of biological molecule.

The generalizations that are needed for the technical control of biological processes will not come from the elaboration either of structural chemistry or of information theory in its conventional sense. These can only provide a background against which effective working concepts can be oriented and rendered more intellectually appealing. The handling of biological material will always be the business of sciences using their own working concepts based essentially on the not very deeply analysed concept of specific pattern with which we have been concerned.

REFERENCES

Ada, G. L. and Perry, B. (1954*a*). Studies on the soluble complement fixing antigens of influenza virus. III. The nature of the antigen. *Aust. J. exp. Biol. med. Sci.* **32**, 179.

Ada, G. L. and Perry, B. T. (1954*b*). The nucleic acid content of influenza virus. *Aust. J. exp. Biol. med. Sci.* **32**, 453.

Ada, G. L. and Perry, B. T. (1955*a*). Infectivity and nucleic acid content of influenza virus. *Nature, Lond.*, **175**, 209.

Ada, G. L. and Perry, B. (1955*b*). Specific differences in the nucleic acids from A and B strains of influenza viruses. *Nature, Lond.*, **175**, 854.

Ada, G. L., Perry, B. T. and Pye, J. (1953). Studies on the soluble complement fixing antigen of influenza virus. II. Serological behaviour of the antigen. *Aust. J. exp. Biol. med. Sci.* **31**, 391.

Allfrey, V., Daly, M. M. and Mirsky, A. E. (1953). Synthesis of protein in the pancreas. II. The role of ribonucleoprotein in protein synthesis. *J. gen. Physiol.* **37**, 157.

Anderson, D., Billingham, R. E., Lampkin, G. H. and Medawar, P. B. (1951). The use of skin grafting to distinguish between monozygotic and dizygotic twins in cattle. *Heredity*, **5**, 379.

Barrett, M. K. and Hansen, W. H. (1953). Resistance to tumour implantation induced by red cell stromata. *Cancer Res.* **13**, 269.

Battisto, J. R. and Chase, M. W. (1955). Immunological paralysis in guineapigs fed allergenic chemicals. *Fed. Proc.* **14**, 456.

Beale, A. J. (1954). A comparison of the soluble antigen production by tissues infected with preparations of extracellular and intracellular influenza virus. *J. Hyg., Camb.*, **52**, 225.

Berenblum, I. and Shubik, P. (1947*a*). The role of croton oil applications associated with a single painting of a carcinogen, in tumour induction of the mouse's skin. *Brit. J. Cancer*, **1**, 379.

Berenblum, I. and Shubik, P. (1947*b*). A new quantitative approach to the study of the stages of chemical carcinogenesis in the mouse's skin. *Brit. J. Cancer*, **1**, 383.

Bernal, J. D. and Carlisle, C. H. (1948). Unit cell measurements of wet and dry crystalline turnip yellow mosaic virus. *Nature, Lond.*, **162**, 139.

BILLINGHAM, R. E., BRENT, L. and MEDAWAR, P. B. (1953). 'Actively acquired tolerance' of foreign cells. *Nature, Lond.*, **172**, 603.

BILLINGHAM, R. E., BRENT, L. and MEDAWAR, P. B. (1955). Acquired tolerance of skin homografts. *Ann. N.Y. Acad. Sci.* **59**, 409.

BJORNEBOE, M., GORMSEN, H. and LUNDQUIST, F. R. (1947). Further experimental studies of the role of plasma cells as antibody producers. *J. Immunol.* **55**, 121.

BLIX, U., ILAND, C. N. and STACEY, M. (1954). The serological activity of desoxypentosenucleic acids. *Brit. J. exp. Path.* **35**, 241.

BOLTON, E. (1954). Biosynthesis of nucleic acid in *Escherichia coli. Proc. nat. Acad. Sci., Wash.*, **49**, 764.

BOOTH, P. B., DUNSFORD, I., GRANT, J. and MURRAY, S. (1953). Hæmolytic disease in first-born infants. *Brit. med. J.* **ii**, 41.

BOULANGER, P. et MONTREUIL, J. (1952). Études sur les acides nucléiques. III. Échange du phosphore radioactif dans les ribonucléotides du foie de rat. *Biochim. biophys. Acta*, **9**, 619.

BOYD, J. K. F. (1950). The symbiotic bacteriophages of *Salmonella typhi-murium. J. Path. Bact.* **62**, 501.

BURNET, F. M. (1952). *Natural History of Infectious Disease.* Cambridge Univ. Press.

BURNET, F. M. (1954*a*). Recent work on the intrinsic qualities of influenza: somatic and genetic aspects. *WHO Bull.* **8**, 661.

BURNET, F. M. (1954*b*). The newer approach to immunity in its bearing on medicine and biology. *Brit. med. J.* **ii**, 189.

BURNET, F. M. and BULL, D. R. (1943). Changes in influenza virus associated with adaptation to passage in chick embryo. *Aust. J. exp. Biol. med. Sci.* **21**, 55.

BURNET, F. M. and EDNEY, M. (1951). Recombinant viruses obtained from double infections with the influenza A viruses MEL and Neuro-WS. *Aust. J. exp. Biol. med Sci.* **29**, 353.

BURNET, F. M. and FENNER, F. (1949). *The Production of Antibodies.* Melbourne: Macmillan.

BURNET, F. M. and LIND, P. E. (1954*a*). Reactivation of heat inactivated influenza virus by recombination. *Aust. J. exp. Biol. med Sci.* **32**, 133.

BURNET, F. M. and LIND, P. E. (1954*b*). Recombination of influenza viruses in the de-embryonated egg. II. The conditions for recombination and the evidence for the possible existence of diploid influenza virus. *Aust. J. exp. Biol. med. Sci.* **32,** 153.

Burnet, F. M. and Lind, P. E. (1954*c*). Genetics of virulence in influenza viruses. *Nature, Lond.*, **173,** 627.

Burnet, F. M. and Lind, P. E. (1954*d*). An analysis of the adaptation of an influenza virus to produce lesions in the mouse lung. *Aust. J. exp. Biol. med. Sci.* **32,** 711.

Burnet, F. M. and Lind, P. E. (1955). Recombination between the influenza virus strains WSE and CAM. *Aust. J. exp. Biol. med. Sci.* **33,** 281.

Burnet, F. M., Lind, P. E. and Stevens, K. M. (1955). Production of incomplete influenza virus in the de-embryonated egg. *Aust. J. exp. Biol. med. Sci.* **33,** 127.

Burnet, F. M. and Stone, J. D. (1945). Further studies on the O-D change in influenza A virus. *Aust. J. exp. Bio. med. Sci.* **23,** 151.

Buxton, A. (1954). Antibody production in avian embryos and young chicks. *J. gen. Microbiol.* **10,** 398.

Cambpell, P. N. and Work, T. S. (1953). Biosynthesis of proteins. *Nature, Lond.*, **171,** 997.

Chase, M. W. (1952). The allergic state. *Bacterial and Mycotic Diseases of Man*, 2nd edition ed. by R. J. Dubos. Philadelphia: Lippincott, p. 168.

Chase, M. W. (1955). Differentiation between roles of white cell in transfer of contact dermatitis and in development of antibody. *Fed. Proc.* **14,** 458.

Clayton, C. C. and Spain, J. D. Jr. (1954). Reticulo-endothelial system and azo dye carcinogenesis. *Fed. Proc.* **13,** 193.

Clinton, R. F., Harington, C. R. and Yuill, M. E. (1938). Studies in synthetic immunochemistry. II. Serological investigation of O-B-glucosidotyrosyl derivates of proteins. *Biochem. J.* **32,** 1111.

Cohen, S. S. and Barner, H. D. (1954). Studies on unbalanced growth in *Escherichia coli. Proc. nat. Acad. Sci., Wash.*, **40,** 885.

Cohen-Bazire, G. et Jolit, M. (1953). Isolement par selection de mutants d'*Escherichia coli* synthétisants spontanément l'amylomaltase et le β-galactosidase. *Ann. Inst. Pasteur*, **84,** 937.

Cohn, M. and Monod, J. (1953). Specific inhibition and induction of enzyme biosynthesis. *Symp. Soc. Gen. Microbiol., Adaptation in Micro-organisms.* Cambridge Univ. Press, p. 132.

REFERENCES

COHN, M. and TORRIANI, A. M. (1953). The relationships in biosynthesis of the β-galactosidase and Pz proteins of *E. coli*. *Biochim. biophys. Acta*, **10**, 280.

COLE, L. R. and FAVOUR, C. B. (1955). Correlations between plasma protein fractions and the passive transfer of delayed and intermediate cutaneous reactivity to tuberculin. *J. exp. Med.* **101**, 391.

COLE, L. J., FISHLER, M. C. and BOND, V. P. (1953). Subcellular fractionation of mouse spleen radiation protection activity. *Proc. nat. Acad. Sci., Wash.*, **39**, 759.

COMMONER, B., YAMADA, M., RODENBERG, S., TUNG-YUE WANG and BASLER, E. (1953). The proteins synthesized in tissue infected with tobacco mosaic virus. *Science*, **118**, 529.

COONS, A. H. (1954). Labelled antigens and antibodies. *Annu. Rev. Microbiol.* **8**, 333.

COONS, A. H. and KAPLAN, M. H. (1950). Localization of antigen in tissue cells. II. Improvements in a method for the detection of antigen by means of fluorescent antibody. *J. exp. Med.* **91**, 1.

COONS, A. H., LEDUC, E. H. and CONNOLLY, J. N. (1953). Immunochemical studies of antibody response in the rabbit. *Fed. Proc.* **12**, 439.

COONS, A. H., LEDUC, E. H. and KAPLAN, M. H. (1951). Localization of antigen in tissue cells. VI. The fate of injected foreign proteins in the mouse. *J. exp. Med.* **93**, 173.

DALGLIESH, C. E. (1953). The template theory and the role of transpeptidation in protein biosynthesis. *Nature, Lond.*, **171**, 1027.

DE DEKEN-GRENSON, M. (1953). Étude de la vitesse de renouvellement du phosphore de l'acide ribonucléique dans des cellules effectuant une synthèse de proteines rapide et directement mesurable. III. Le pancréas de souris. *Biochim. biophys. Acta*, **10**, 480.

DELBRUCK, M. (1954). On the replication of desoxyribonucleic acid (DNA). *Proc. nat. Acad. Sci., Wash.*, **40**, 783.

DEPOUX, R. and ISAACS, A. (1954). Interference between influenza and vaccinia viruses. *Brit. J. exp. Path.* **35**, 415.

DETTWILER, H. A., HUDSON, N. P. and WOOLPERT, O. C. (1940). The comparative susceptibility of fetal and postnatal guinea-pigs to the virus of epidemic influenza. *J. exp. Med.* **72**, 623.

DIXON, F. J., BUKANTZ, S. C., DAMMIN, G. J. and TALMAGE, D. W. (1951). Fate of I^{131} labelled bovine γ-globulin in rabbits. *Fed. Proc.* **10**, 553.

DIXON, F. J., BUKANTZ, S. C., DAMMIN, G. J. and TALMAGE, D. W. (1953). Fate of I^{131} labelled bovine gamma globulin in rabbits, in *The Nature and Significance of the Antibody Response*, ed. A. M. Pappenheimer, Jr. Symp. N.Y. Acad. Med. No. 5, New York: Columbia University Press, p. 170.

DIXON, F. J. and MAURER, P. H. (1955). Immunologic unresponsiveness induced by protein antigens. *J. exp. Med.* **101**, 245.

DIXON, F. J., MAURER, P. H. and DEICHMILLER, M. P. (1954). Primary and specific anamnestic antibody responses of rabbits to heterologous serum protein antigens. *J. Immunol.* **72**, 179.

DONALD, H. B. and ISAACS, A. (1954). Counts of influenza virus particles. *J. gen. Microbiol.* **10**, 457.

DOUNCE, A. L. (1952). Duplicating mechanism for peptide chain and nucleic acid synthesis. *Enzymologia*, **15**, 251.

DUNSFORD, I., BOWLEY, C. C., HUTCHINSON, A. M., THOMPSON, J. S., SANGER, R. and RACE, R. R. (1953). A human blood group chimæra. *Brit. med. J.* **ii**, 81.

EARLE, W. R. and NETTLESHIP, A. (1943). Production of malignancy *in vitro*. V. Results of injections of cultures into mice. *J. nat. Cancer Inst.* **4**, 213.

ELSON, D. and CHARGAFF, E. (1954). Regularities in the composition of pentose nucleic acids. *Nature, Lond.*, **173**, 1037.

FAGRAEUS, A. (1948). Antibody production in relation to the development of plasma cells. *Acta med. scand.* Supp. 204.

FAZEKAS DE ST GROTH, S. and CAIRNS, H. J. F. (1952). Quantitative aspects of influenza virus multiplication. IV. Definition of constants and general discussion. *J. Immunol.* **69**, 173.

FAZEKAS DE ST GROTH, S. and EDNEY, M. (1952). Quantitative aspects of influenza virus multiplication. II. Heterologous interference. *J. Immunol.* **69**, 160.

FAZEKAS DE ST GROTH, S. and GRAHAM, D. M. (1954). The production of incomplete virus particles among influenza strains: experiments in eggs. *Brit. J. exp. Path.* **35**, 60.

FELTON, L. D. (1949). The significance of antigen in animal tissues. *J. Immunol.* **61**, 107.

Fox, A. S. (1954). Protein synthesis and genetics. *Nature, Lond.*, **173**, 350.

Franklin, R. E. (1955). Structure of tobacco mosaic virus. *Nature, Lond.*, **175**, 379.

Franklin, R. E. and Commoner, B. (1955). X-ray diffraction by an abnormal protein B8 associated with tobacco mosaic virus. *Nature, Lond.*, **175**, 1076.

Fraser, K. B. (1953). Genetic interaction and interference between the MEL and NWS of influenza A virus. *Brit. J. exp. Path.* **34**, 319.

Fruton, J. S. (1950). The rôle of proteolytic enzymes in the biosynthesis of peptide bonds. *Yale J. Biol. Med.* **22**, 263.

Gale, E. G. and Folkes, J. P. (1954). Effect of nucleic acids on protein synthesis and amino-acid incorporation in disrupted staphylococcal cells. *Nature, Lond.*, **173**, 1223.

Gale, E. G. and Folkes, J. P. (1955). Promotion of incorporation of amino-acids by specific di- and trinucleotides. *Nature, Lond.*, **175**, 592.

Gamow, G. and Metropolis, N. (1954). Numerology of peptide chains. *Science*, **120**, 779.

Gard, S. and Magnus, P. von (1946). Studies on interference in experimental influenza. II. Purification and centrifugation experiments. *Ark. Kemi Min. Geol.* 24B, No. 8, I.

Gell, P. G. H. (1944). Sensitization to 'tetryl'. *Brit. J. exp. Path.* **25**, 174.

Gell, P. G. H., Harington, C. R. and Pitt-Rivers, R. (1946). The antigenic function of simple chemical compounds, production of precipitins in rabbits. *Brit. J. exp. Path.* **27**, 267.

Gitlin, D., Landing, B. H. and Whipple, A. (1953). The localization of homologous plasma proteins in the tissues of young human beings as demonstrated with fluorescent antibodies. *J. exp. Med.* **97**, 163.

Gorbman, A. (1949). Tumorous growths in the pituitary and trachea following radiotoxic dosages of I^{131}. *Proc. Soc. Exp. Biol., N.Y.*, **71**, 237.

Gottschalk, A. (1953). On the mechanism of enzyme action. *Rev. of pure and app. Chem.* **3**, 179.

Gotlieb, T. and Hirst, G. K. (1954). The experimental production of combination forms of virus. III. The formation of doubly antigenic particles from influenza A and B virus and a study of the ability of individual particles of X virus to yield two separate strains. *J. exp. Med.* **99**, 307.

Green, H. N. (1954). An immunological concept of cancer: a preliminary report. *Brit. med. J.* ii, 1374.

Haddow, A. (1938). Cellular inhibition and origin of cancer. *Acta de l'Union intern. contre Cancer*, **3**, 342.

Haddow, A. (1953). The chemical and genetic mechanisms of carcinogenesis, in *The Physiopathology of Cancer*, ed. by F. Homburger and W. H. Fishman. London: Cassell, p. 441.

Haldane, J. B. S. (1954). *The Biochemistry of Genetics.* London: Allen & Unwin.

Hall, C. E., Hall, O. and Cunningham, A. W. B. (1953). Cited by Green, 1954.

Hanan, R. and Oyama, J. (1954). Inhibition of antibody formation in mature rabbits by contact with the antigen at an early age. *J. Immunol.* **73**, 49.

Harris, S., Harris, T. N. and Farber, M. B. (1954*a*). Studies on the transfer of lymph node cells. I. Appearance of antibody in recipients of cells from donor rabbits injected with antigen. *J. Immunol.* **72**, 148.

Harris, T. N., Harris, S. and Farber, M. B. (1954*b*). Transfer to X-irradiated rabbits of lymph node cells incubated in vitro with *Shigella paradysenteriæ*. *Proc. Soc. exp. Biol., N.Y.*, **86**, 549.

Haurowitz, F. (1952*a*). *The Chemistry and Biology of Proteins.* New York: Academic Press.

Haurowitz, F. (1952*b*). The mechanism of the immunological response. *Biol. Rev.* **27**, 247.

Haurowitz, F. and Crampton, C. F. (1952). The fate in rabbits of intravenously injected I^{131} endovoalbumin. *J. Immunol.* **68**, 73.

Haxthausen, H. (1943). The pathogenesis of allergic eczema elucidated by transplantation experiments on identical twins. *Acta derm.-venereol., Stockh.*, **23**, 438.

Hayes, S. P. and Dougherty, T. F. (1954). Studies on local antibody production: demonstration of agglutination by lymphocytes. *J. Immunol.* **73**, 95.

Henle, W., Henle, G. and Rosenberg, E. G. (1947). The demonstration of one step growth curves of influenza virus through the blocking effect of irradiated virus on further infection. *J. exp. Med.* **86**, 423.

Hershey, A. F. and Chase, M. (1952). Independent functions of viral proteins and nucleic acid in growth of bacteriophage. *J. gen. Physiol.* **36**, 39.

Hirst, G. K. and Gotlieb, T. (1953). The experimental production of combination forms of virus. I. Occurrence of combination forms after simultaneous inoculation of the allantoic sac with two distinct strains of influenza virus. *J. exp. Med.* **98**, 41.

Hokin, L. E. and Hokin, M. R. (1954). The incorporation of ^{32}P into the nucleotides of RNA in pancreas slices during enzyme synthesis and secretion. *Biochim. biophys. Acta*, **13**, 401.

Holt, L. B. (1949). Quantitative studies in diphtheria prophylaxis: the primary response. *Brit. J. exp. Path.* **30**, 289.

Horowitz, N. H. and Owen, R. D. (1954). Physiological aspects of genetics. *Annu. Rev. Physiol.* **16**, 81.

Horsfall, F. L. Jr. (1954). On the reproduction of influenza virus: quantitative studies with procedures which enumerate infective and hæmagglutinating virus particles. *J. exp. Med.* **100**, 135.

Hotchkiss, R. D. and Marmur, J. (1954). Double marker transformations as evidence of linked factors in deoxyribonucleate transforming agents. *Proc. nat. Acad. Sci., Wash.*, **40**, 55.

Hoyle, L. (1953). The multiplication of the influenza virus considered in relation to the general problem of virological multiplication. *Symp. Soc. Gen. Microbiol., Nature of Virus Multiplication*, Cambridge Univ. Press, p. 225.

Hoyle, L., Jolles, B. and Mitchell, R. G. (1954). The incorporation of radioactive phosphorus in the influenza virus and its distribution in serologically active virus fractions. *J. Hyg., Camb.*, **52**, 119.

Ingraham, J. A. (1951). Artificial radioactive antigens. II. The metabolism of S^{35} sulfanilic acid-azo-bovine-γ-globulin in normal and immune mice. *J. infect. Dis.* **89**, 117.

Isaacs, A. and Fulton, F. (1953). Interference in the chick chorion. *J. gen. Microbiol.* **9**, 132.

Janeway, C. A., Apt, L. and Gitlin, D. (1953). Agammaglobulinemia. *Trans. Ass. Amer. Physns.* **66**, 200.

Kaplan, H. S. (1954). On the etiology and pathogenesis of the leukemias: a review. *Cancer Res.* **14**, 535.

KAPLAN, J. G. (1955). The alteration of intracellular enzymes. I. Yeast catalase and the Euter effect. *Exp. Cell Res.* **8**, 305.

KERR, W. R. and ROBERTSON, M. (1954). Passively and actively acquired antibodies for *Trichomonas fœtus* in very young calves. *J. Hyg., Camb.*, **52**, 253.

KNIGHT, C. A. (1946). The precipitin reactions of highly purified influenza viruses and related materials. *J. exp. Med.* **83**, 281.

KOPROWSKI, H. (1955). Actively acquired tolerance to a mouse tumour. *Nature, Lond.*, **175**, 1087.

KORITZ, S. B. and CHANTRENNE, H. (1954). The relationship of ribonucleic acid to the *in vitro* incorporation of radioactive glycine into the proteins of reticulocytes. *Biochim. biophys. Acta*, **13**, 209.

KRUSE, H. and MCMASTER, P. D. (1949). The distribution and storage of blue antigenic azoproteins in the tissues of mice. *J. exp. Med.* **90**, 425.

KUHNS, W. J. and PAPPENHEIMER, A. M., Jr. (1952). Immunochemical studies of antitoxin produced in normal and allergic individuals hyperimmunized with diphtheria toxoid. I. Relationship of skin sensitivity to purified diphtheria toxoid to the presence of circulating nonprecipitating antitoxin. *J. exp. Med.* **95**, 363.

LAWS, J. O. and WRIGHT, G. P. (1952). The disposal and organ-distribution of radio-iodinated bovine serum proteins in control and specifically sensitized rabbits. *Brit. J. exp. Path.* **33**, 343.

LEYON, H. (1951). Some physico-chemical properties of spontaneous mouse encephalomyelitis virus strain FA. *Exp. Cell Res.* **2**, 207.

LIND, P. E. and BURNET, F. M. (1953). Back-recombination of influenza A strains obtained in recombination experiments. *Aust. J. exp. Biol. med. Sci.* **31**, 361.

LIND, P. E. and BURNET, F. M. (1954). Recombination between neurotropic and non-neurotropic influenza virus strains. *Aust. J. exp. Biol. med. Sci.* **32**, 437.

LITTLE, C. C. (1914). A possible mendelian explanation for a type of inheritance apparently non-mendelian in nature. *Science*, **40**, 904.

LIU, O. C., BLANK, H., SPIZIZEN, J. and HENLE, W. (1955). The incorporation of radioactive phosphorus into influenza virus. *J. Immunol.* **74**, 415.

LOFTFIELD, R. B., GROVER, J. W. and STEPHENSON, M. L. (1953). Possible role of proteolytic enzymes in protein synthesis. *Nature, Lond.*, **171**, 1024.

VON MAGNUS, P. (1952). Propagation of the PR8 strain of influenza A virus in chick embryos. IV. Studies on the factors involved in the formation of incomplete virus upon serial passage of undiluted virus. *Acta Path. et Microbiol. scand.* **30**, 311.

MANDELSTAM, J. and YUDKIN, J. (1952). Studies in biochemical adaptation: some aspects of galactozymase production by yeast in relation to the 'mass action' theory of enzyme adaptation. *Biochem. J.* **51**, 686.

MANSON, E. E. D., POLLOCK, M. R. and TRIDGELL, E. J. (1954). A comparison of the properties of penicillinase produced by *B. subtilis* and *B. cereus* with and without addition of penicillin. *J. gen. Microbiol.* **11**, 493.

MANSON, L. A. (1953). The metabolism of ribonucleic acid in normal and bacteriophage infected *Escherichia coli*. *J. Bact.* **66**, 703.

MARKHAM, R. (1953). Nucleic acids in virus multiplication. *Symp. Soc. Gen. Microbiol., Nature of Virus Multiplication*, Cambridge Univ. Press, p. 85.

MARKHAM, R. and SMITH, K. M. (1949). Studies on the virus of turnip yellow mosaic. *Parasitology*, **39**, 330.

MAURER, P. H., DIXON, F. J. and TALMADGE, D. W. (1953). A quantitative comparison of antibody produced by X-radiated and normal rabbits. *Proc. Soc. exp. Biol., N.Y.*, **83**, 163.

MEDAWAR, P. B. (1947). Cellular inheritance and transformation. *Biol Rev.* **22**, 360.

MILLER, E. C. and MILLER, J. A. (1952). *In vivo* combinations between carcinogens and tissue constituents and their possible role in carcinogenesis. *Cancer Res.* **12**, 547.

MILLER, J. A. and MILLER, E. C. (1948). The carcinogenicity of certain derivatives of *p*-dimethyl-aminoazobenzene in the rat. *J. exp. Med.* **87**, 139.

MILLER, L. L., BLY, C. G. and BALE, W. F. (1954). Plasma and tissue proteins produced by non-hepatic rat organs as studied with lysine-C^{14}. *J. Exp. Med.* **99**, 133.

MOGABGAB, W. J., GREEN, I. J. and KIERKHISING, O. C. (1954). Primary isolation and propagation of influenza virus in cultures of human embryonic renal tissue. *Science*, **120**, 320.

MONOD, J., COHEN-BAZIRE, G. et COHN, W. (1951). Sur la biosynthèse de la β-galactosidase chez *E. coli*. La spécificité de l'induction. *Biochim. biophys. Acta*, **7**, 585.

MONOD, J. and COHEN-BAZIRE, G. (1953), cited in Cohn and Monod (1953).

MONOD, J. et COHN, M. (1952). La biosynthèse induite des enzymes (adaptation enzymatique). *Advanc. Enzymol.* **13**, 67.

MONOD, J., PAPPENHEIMER, A. M., Jr., et COHEN-BAZIRE, G. (1952). La cinétique de la biosynthèse de la β-galactosidase chez *E. coli* considérée comme fonction de la croissance. *Biochim. biophys. Acta*, **9**, 648.

OAKLEY, C. L., BATTY, I. and WARRACK, G. H. (1951). Local production of antibodies. *J. Path. Bact.* **63**, 33.

OWEN, R. D. (1945). Immunogenetic consequences of vascular anastomoses between bovine twins. *Science*, **102**, 400.

OWEN, R. D., WOOD, H. R., FOORD, A. G., STURGEON, P. and BALDWIN, L. G. (1954). Evidence for actively acquired tolerance to *Rh* antigens. *Proc. nat. Acad. Sci., Wash.*, **40**, 420.

PANUM, P. L. (1847). Beobachtungen uber das Maserncontagium. *Virchows Arch.* **1**, 492. Also in *Medical Classics*, **3**, 829 (1939).

PARDEE, A. B. (1954). Nucleic acid precursors and protein synthesis. *Proc. nat. Acad. Sci., Wash.*, **40**, 263.

PAULING, L. (1948). Nature of forces between large molecules of biological interest. *Nature, Lond.*, **161**, 707.

PERRY, B. T. and BURNET, F. M. (1953). Recombination studies with two influenza virus B strains. *Aust. J. exp. Biol. med. Sci.* **31**, 519.

PERRY, B. T., VAN DEN ENDE, M. and BURNET, F. M. (1954). Recombination with two influenza B strains in the de-embryonated egg. *Aust. J. exp. Biol. med. Sci.* **32**, 469.

POLLARD, M. and RUSSELL, R. H. (1954). Specificity of antineoplastic phenomenon induced in mice by carcinogenic agents. *Proc. Soc. exp. Biol., N.Y.*, **86**, 186.

POLLOCK, M. R. (1950). Penicillinase adaptation in *B. cereus*: adaptive enzyme formation in the absence of free substrate. *Brit. J. exp. Path.* **31**, 739.

POLLOCK, M. R. (1953). Stages in enzyme adaptation in mucoproteins. *Symp. Soc. for Gen. Microbiology, Adaptation in Micro-organisms*. Cambridge Univ. Press, p. 150.

POLLOCK, M. R. and PERRET, C. J. (1951). The relation between fixation of penicillin sulphur and penicillinase adaptation in *B. cereus. Brit. J. exp. Path.* **32**, 387.

PURVES, H. D. and GRIESBACH, W. E. (1946). Studies in experimental goitre. VII. Thyroid carcinomata in rats treated with thiourea. *Brit. J. exp. Path.* **27**, 294.

QUASTLER, H. (1953). Ed. *Information Theory in Biology*, Urbana.

RAFFEL, S. and FORNEY, J. E. (1948). The role of the 'wax' of the tubercle bacillus in establishing delayed hypersensitivity. I. Hypersensitivity to a simple chemical substance, picryl chloride. *J. exp. Med.* **88**, 485.

REISS, E., MERTENS, E. and EHRICH, W. E. (1950). Agglutination of bacteria by lymphoid cells *in vitro. Proc. Soc. exp. Biol., N.Y.*, **74**, 732.

RICH, A., DUNITZ, J. D. and NEWMARK, P. (1955). Structure of polymerized tobacco plant protein and tobacco mosaic virus. *Nature, Lond.*, **175**, 1074.

SCHAFFER, F. L. and SCHWERDT, C. E. (1955). Nucleic acid composition of purified preparations of polio virus. *Fed. Proc.* **14**, 275.

SCHINCKEL, P. G. and FERGUSON, K. A. (1953). Skin transplantation in the fœtal lamb. *Aust. J. of Biol. med. Sci.* **6**, 533.

SCHRAMM, G., SCHUMACHER, G. and ZILLIG, W. (1955). An infectious nucleoprotein from tobacco mosaic virus. *Nature, Lond.*, **175**, 549.

SCHRÖDINGER, E. (1944). *What is Life?* Cambridge Univ. Press.

SCHWARTZ, D. (1955). Speculations on gene action and protein specificity. *Proc. nat. Acad. Sci., Wash.*, **41**, 300.

SIMON, F. A., SIMON, M. G., RACKEMANN, F. M. and DIENES, L. (1934). The sensitization of guinea pigs to poison ivy. *J. Immunol.* **27**, 113.

SNELL, G. D. (1953). Transplantable tumours, in *The Physiopathology of Cancer*, ed. by F. HOMBURGER and W. H. FISHMAN. London: Cassell, p. 338.

SPIEGELMAN, S. and HALVORSON, H. O. (1953). The nature of the precursor on the induced synthesis of enzymes. *Symp. Soc. Gen. Microbiol., Adaptation in Micro-organisms.* Cambridge Univ. Press, p. 98.

STARK, D. K. (1955). Studies on pneumococcal polysaccharide. II. Mechanism involved in immunological paralysis. *J. Immunol.* **74**, 130.

STEVENS, K. M. (1953). Antigen retention in the rabbit. *J. exp. Med.* **97**, 247.

STEVENS, K. M. (1954). The effect of desoxyribonucleic acid inhibitors upon the replication of influenza virus. *Aust. J. exp. Biol. med Sci.* **32**, 187.

STONE, J. D. (1951). Adsorptive and enzymic behaviour of influenza viruses in the O-D change. *Brit. J. exp. Path.* **32**, 367.

STORMONT, C., WEIR, W. C. and LANE, L. L. (1953). Erythrocyte mosaicism in a pair of sheep twins. *Science*, **118**, 695.

STRONG, L. C. (1926). Changes in the reaction potential of a transplantable tumour. *J. exp. Med.* **43**, 713.

SWENSON, P. A. and GIESE, A. C. (1950). Photoreactivation of galactozymase formation in yeast. *J. cell. comp. Physiol.* **36**, 369.

TALIAFERRO, W. H. and TALIAFERRO, L. G. (1951). The role of the spleen in hæmolysin production in rabbits receiving multiple antigen injections. *J. infect. Dis.* **89**, 143.

TALIAFERRO, W. H. and TALIAFERRO, L. G. (1954). Further studies on the radiosensitive stages of hæmolysin formation. *J. infect. Dis.* **95**, 134.

TALIAFERRO, W. H., TALIAFERRO, L. G. and JANSSEN, E. F. (1952). The localization of X-ray injury to the initial phases of antibody response. *J. infect. Dis.* **91**, 105.

TALMAGE, D. W., DIXON, F. J., BUKANTZ, S. C. and DAMMIN, F. J. (1951). Antigen elimination from the blood as an early manifestation of the immune response. *J. Immunol.* **67**, 243.

TAYLOR, A. R., SHARP, D. G., BEARD, D. and BEARD, J. W. (1943). Isolation and properties of the equine encephalomyelitis (eastern strain). *J. infect. Dis.* **72**, 31.

TODD, A. R. (1954). Chemical structure of the nucleic acids. *Proc. nat. Acad. Sci., Wash.*, **40**, 748.

TOMLIN, S. G. and CALLAN, H. G. (1951). Preliminary account of an electron microscope study of chromosomes from newt oocytes. *Quart. J. micr. Sci.* **92**, 221 (cited by Haldane, 1954).

UHLER, M. and GARD, S. (1954). Lipid content of 'standard' and 'incomplete' influenza A virus. *Nature, Lond.*, **173**, 1041.

VIGNEAUD, V. DU, RESSLER, C., SWAN, J. M., ROBERTS, C. W., KATSOYANNIS, P. G. and GORDON, S. (1953). The synthesis of an octapeptide amide with hormonal activity of oxytocin. *J. Amer. chem. Soc.* **75**, 4879.

VILCHES, A. and HIRST, G. K. (1947). Interference between neurotropic and other unrelated viruses. *J. Immunol.* **57**, 125.

WATSON, B. K. and COONS, A. H. (1954). Studies of influenza virus infection in the chick embryo using fluorescent antibody. *J. exp. Med.* **99**, 419.

WATSON, J. D. and CRICK, F. H. C. (1953). Molecular structure of nucleic acids. *Nature, Lond.*, **171**, 737.

WEILER, E. (1952). Die Anderung der serologischen Organspezifität beim Buttergelb-Tumor der Ratte im Vergleich zu normaler Leber. *Z. Naturf.* **7**, 324.

WEISS, P. (1947). The problem of specificity in growth and development. *Yale J. Biol. Med.* **19**, 235.

WEISS, P. (1950). Perspectives in the field of morphogenesis. *Quart. Rev. Biol.* **25**, 177.

WERNER, G. H. and SCHLESINGER, R. W. (1954). Morphological and quantitative comparison between infectious and non-infectious forms of influenza virus. *J. exp. Med.* **100**, 203.

WESSLEN, T. (1952). Studies on the role of lymphocytes in antibody production. *Acta derm.-venereol., Stockh.* **32**, 265.

WHITE, R. G., COONS, A. H. and CONNOLLY, J. N. (1953). Cellular morphology of antibody production. *Fed. Proc.* **12**, 445.

WOLF, G. (1952). *Chemical Induction of Cancer*. London: Cassell.

WRIGHT, G. P. (1955). Botulinum and tetanus toxins. *Symp. Gen. Microbiol., Mechanisms of Microbial Pathogenicity*, Cambridge Univ. Press, p. 78.

YOUNG, J. Z. (1954). Memory, heredity and information, in *Evolution as a Process*; ed. by J. HUXLEY, A. C. HARDY and E. B. FORD. London: Allen & Unwin, p. 281.

YUDKIN, J. (1938). Enzyme variation in micro-organisms. *Biol. Rev.* **13**, 93.

ZINDER, N. D. and LEDERBERG, J. (1952). Genetic exchange in Salmonella. *J. Bact.* **64**, 679.

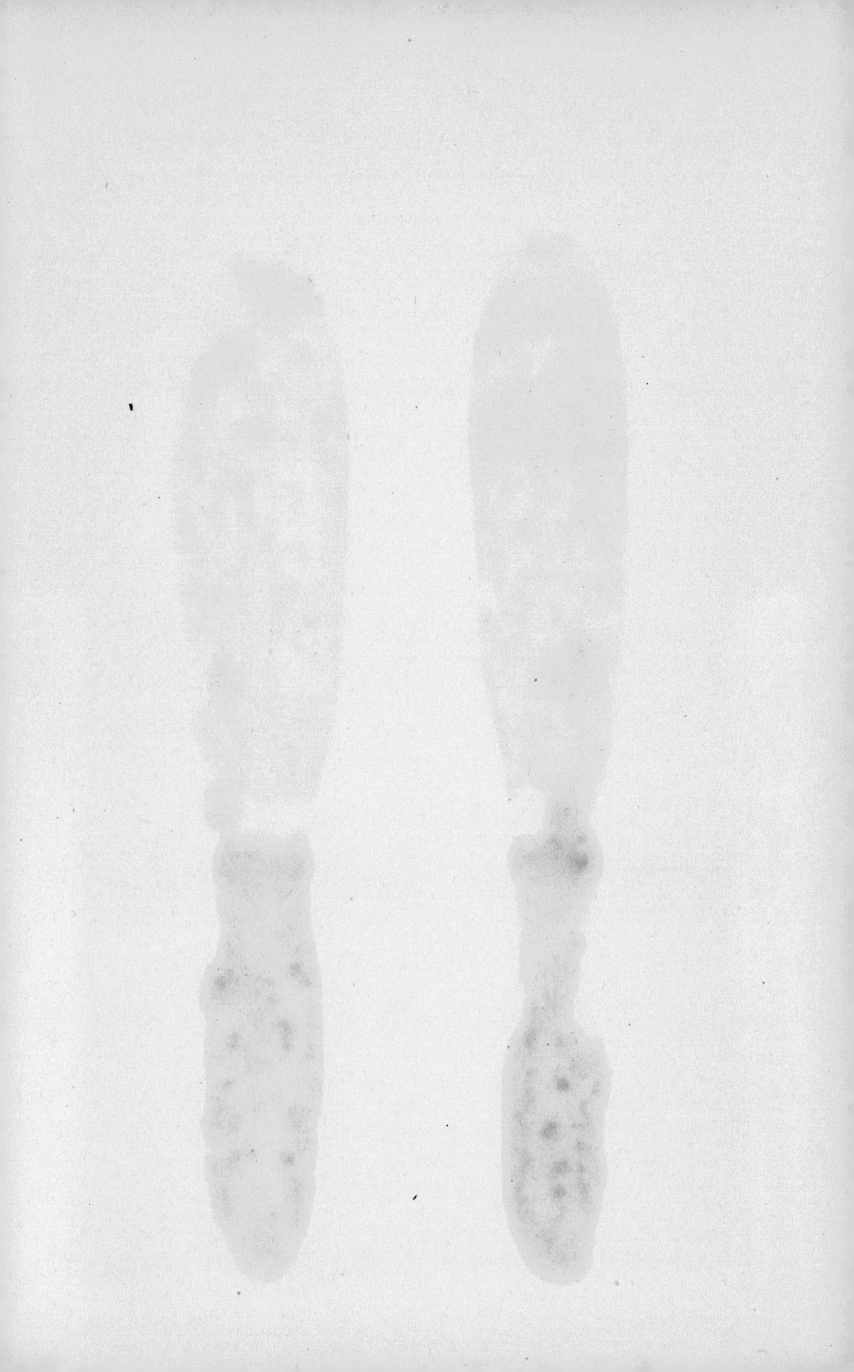